Construction Claims Manual for Residential Contractors

Jonathan F. Hutchings

McGraw-Hill

New York San Francisco Washington, D.C. Auckland Bogota
Caracas Lisbon London Madrid Mexico City Milan
Montreal New Delhi San Juan Singapore
Sydney Tokyo Toronto

McGraw-Hill

A Division of The McGraw-Hill Companies

1 2 3 4 5 6 7 8 9 0 DOC/DOC 9 0 3 2 1 0 9 8

ISBN 0-07-031838-7 [hardcover]

ISBN 0-07-031839-5 [paperback]

The sponsoring editor for this book was Zoe G. Foundotos, the editing supervisor was Sally Anne Glover, and the production supervisor was Pamela Pelton.

Printed and bound by R. R. Donnelley & Sons Company.

This book is printed on recycled, acid-free paper containing a minimum of 50% recycled, de-inked fiber.

Contents

Foreword

I've known the author of this book for 20 years and he knows as well as I that the American construction industry has changed dramatically in these last two decades—in both project execution and complexity of contract documents. The hard lesson learned by owners and contractors alike in the recessions of the 70s and 80s was that increases in the cost and complications of doing business in the construction business meant that <u>all</u> costs must be tightly controlled to preserve profits. Absorbed costs or contingency costs, once the catch-all pigeon hole of residential contractors, suddenly meant immediate loss of profit due to tighter profit margins with no room for overruns or error.

Taking this tactic one step further means not only obtaining a sophistication of the sources of change orders and construction claims for damages, but using that knowledge advantageously in contractual obligations and project operations. In construction management, this is called *claims consciousness*. Claims consciousness begins with the knowledge that even seemingly insignificant problems can escalate through neglect or mismanagement to sizable liquidated damages. Delay of project schedule alone can bankrupt a residential contractor, let alone fixing the problem out of pocket. And as the complexity of today's contracts and projects increases, so too does the potential of problems leading to change orders and construction claims for damages.

The next step in claims consciousness is educating oneself to understand where change orders and claims originate. Steps can then be taken to analyze and alleviate their impact before they grow into financial alligators. The final step in claims consciousness is preparation and prevention of situations leading to change orders and construction claims. This is done by careful analysis of errors and omissions, schedule delay, schedule acceleration, interference, and latent conditions.

From such a contractual position of strength, the disputed parties in construction claims clearly see the advantage in a quick settlement as opposed to a protracted litigation. No one's going to win big there but the lawyers. This book is a must have for all residential builders to develop coordinated systems for change orders and

construction claims prevention. The author has given analysis of those construction claims that have historically been points of contention that apply to every residential construction company. You'll learn what to do about both generating and defending change orders and claims.

This book will help you learn how to make the many complex and diverse methods of claims management more easily understood. What you will learn, design, and develop from this book today is real-world application stuff that you can use on the job tomorrow.

Builders these days know that, to remain competitive, they must learn to apply change order and claims management in a comprehensive and productive real-world manner. In modern business, nothing less will last. Next-step building projects using new materials and execution technologies are rushing at today's builders at breakneck speed. Without a solid, future-minded claims management program, my business would fail within the quarter. Nowadays, project management involves residential builders being sophisticated in all phases of the business, including change orders. Builders are getting used to running the details of *multiple jobs* simultaneously. Decisions relating to expenditures of thousands of dollars are made instantly, every hour of the day. The old sofa at the end wall of the office is not for sitting on, it's for cat-napping on when the action starts going round the clock as project completion dates near.

It's easy for a single change order on a jobsite somewhere to get lost in all this. And when a change order falls through the cracks in project operations, it equates to lost profit. Prior planning prevents loss of profit and keeps you in business. Change order discipline and claims management makes sense in every way for the American builder.

Viable claims management programs are the way we professional builders manage today's construction execution. And one of the best pros I know of, is the author of this book. This edition will definitely be on my shelf and if you're serious about carving out a successful career in residential construction for yourself, my advice to you is to grab this very special book in a hurry before they sell out.

Chad Simmons, President
SJB Construction Corporation
Silicon Valley, California

Preface

My editor asked me to write this book for the typical American builder. After I got off the phone, it occurred to me that although I've got twenty years in this industry I really couldn't define who the typical American builder is. That's because by looking at all the builders I know, there really is no such thing as a typical American builder. Builders are normal, everyday people just like you and me. You can't pick them out at the mall, they look just like everyone else. Because they *are* everyone else. Your neighbor, some guy on the golf course, a woman next to you in the frozen foods section at the market, a husband and wife at the PTA meeting, your uncle. You sure can't tell by looking. Most successful general contractors these days carry nailbags with tools, and briefcases with cellular phones. The typical American builder can't be stereotyped to any common denominator. However, in my research I did find national surveys that found these results:

The "typical" American builder is an entrepreneur who builds fewer than 25 houses a year, has an annual volume of less than $1 million, builds primarily single-family move-up houses of 1,801 to 2,500 square feet, employs one to three full-time staffers, dabbles in remodeling, and has been in the business for more than 10 years. It's someone who works 51 to 60 hours per week, drives a pickup truck, likes to fish and play golf, listens to oldies music at work, and reads six or fewer books per year. (Hopefully, this is one of them).

Today's "typical" American builder is conservative, leans to the right in voting, runs a local hometown business that is lean, with three or less full-time employees. Six out of ten build in one market only, but many have diversified into other businesses: Ninety percent do some remodeling, one in three develops land, and one in four does commercial work on the side. And builders, like people everywhere who work too hard, are always reassessing their priorities. When asked about their most important personal goal between now and the end of the century, spending more time with family topped the list. A close second was staying in business. Sound familiar? Well, that's my objective from here on out. To help you make more profit from the job and stay in business by developing and applying modern change order and claims management techniques.

Profitable residential job practices are looking to modern construction contract management for innovative techniques and project cost savings. The new age of technology has brought in new types of building materials, new project execution technology, and powerful new computer systems using dedicated software programs, opening the way for construction contracts to become increasingly more complex. In a world of the constantly changing market environments of construction the 21st Century, the contractor's only hope of survival are the practices, strategies, and defensive techniques that outlined in this book and have proven themselves to be the most efficient in claims management.

On any construction project of any size, it is important to understand that changes will definitely occur. Therefore, change orders are a natural part of the construction process. Construction loan funders and project financiers inherently understand this fact, and in the project's construction loan provide a contingency fund expressly to accommodate potential changes. This contingency is most often on the order of 10% of the total project cost. Change orders are accepted by lenders as being a common part of doing business in the construction industry. However, there is a large discrepancy between the acceptance and management of this fact by the construction business and the haphazard way residential contractors manage their change orders—often losing money in the process.

Total change order costs (and their impact) encompasses the costs of actual work performed, costs related to any change in method of construction, costs related to the effect of changed work on contract work, and additionally, other impact or inference costs such as delay, schedule acceleration, etc. The primary difficulty lies in the fact that, in most cases, the owner will acknowledge and authorize payment for the cost of actual work performed only and possibly some related expenses. The other costs as described are likely to be left to be absorbed or to be gotten through litigation.

The real talent in change order resolution lies in the ability of the change order preparer and negotiator to move the owner to accept and authorize those consequential cost items that have traditionally been rejected on their initial presentation, before they go to the construction claims stage.

The modern construction industry is so competitive one must be aggressive in the arena of performance of contract documents to be successful. Over the past decade this industry has seen a sharp increase in cost of labor and materials, overhead, and time constraints on project execution. Paralleling this trend has been an increase in the amount of construction law and contract documents. This squeeze resulted in profitable production for today's builder dependent on an uninterrupted project schedule.

Contingencies are no longer a viable part of any bid. Things are just too tight. In the past, what the architect would have previously clarified when issued an RFI (request for information) or RFC (request for clarification), the owner will now process and pay for as a change order to avoid project schedule delay.

On the other hand, cost over runs cannot be absorbed by the contractor any more because the tight profit margin on the bid will simply not allow absorption of any extra cost without direct reciprocal profit loss. Though change orders are a normal part of the construction process, any deviation in a modern construction project schedule can have very large liquidated damages consequences. This has lead to change orders (sometimes referred to as *contract modifications*) becoming the harbinger of the construction industry. Problems in the areas of change orders and construction claims if not arbitrated successfully invariably lead to litigation. The bid-getting strategy of "low-balling" (going in under real cost and making up the difference in extras and change orders) was also a factor in bringing the change order and construction claim to a legal tactic state.

Construction litigation has exploded in the last decade and is now the fourth largest cause of lawsuits in the nation. It has become a multibillion dollar business in America. And it is merciless. This industry is now comprised of high-stakes profit, high-risk financing, a diverse spectrum of highly skilled lawyers who specialize in construction litigation, and thousands of professional residential builders and contractors trying to make a decent living. (Needless to say, it's a volatile mix that detonates regularly.)

Construction claims are the legal instruments that enforce payment in legal open court for modern construction project disputes between the vested parties. They are used as such not only in the construction industry, but also by the insurance industry for repair claims adjustment compensations, and by the legal industry in court litigation. This book is not geared towards those arenas. This book is written with one purpose in mind: providing residential builders and contractors a broader practical application knowledge of change orders and construction claims, and an understanding of the business scope, meaning, defense, and differing applications of their use. It is essential that modern residential builders understand where their contractual responsibilities end and where their rights to get paid for additional work or delay begin.

Accordingly, the design of this book is not to reprint volumes of the many different construction law code regulations and findings, but to serve as a comprehensive reference book of the specific topics and use of change orders and construction claims for residential builders and contractors. It does not encompass the entirety of possible contractual scenarios. (It's a perfect world, right?)

The book's chapters cover offensive and defensive use of change orders, legal ramifications of production schedule delay, related filing schedules, and legal notices—all of which are necessary to successfully use change orders and claims without being on the losing end of liquidated damages resulting from project delay. Covering residential construction claims, this book is unique because it is not broadly esoteric in nature but specifically addresses the everyday needs of the residential builder or contractor searching for construction claims profit versus loss advice from an industry expert. As an added reader benefit, this book provides tips on where to find all those hidden items in the contract that will maximize the builder's total compensation in pursuing construction claims, how to effectively prepare and present claims for damages, and strategies for generating/defending them.

This book also addresses the confusion, expense, and worry of project change orders and construction claims for the residential builder's concerns as never before. The tips and techniques learned from this book can be applied by any builder to produce more efficient performance from contract documents. This book will show you how to analyze contracts to uncover all those effects on your projects for which you are entitled to get paid, maximize your position in the hardball arena of contract negotiation, consistently maximize your progress payments, and to improve your win record in arbitration and litigation.

Remember, in matters of real-world applications of construction contracts always research the contract specifications for code compliance thoroughly and consult an attorney in all contractual obligation matters. This book does not offer legal advice, nor should it be construed as such. Further, please be informed that the facts, subjects, references, and standards contained herein are as accurate as this time of writing will allow.

Respectfully submitted,

Jonathan F. Hutchings

Acknowledgments

The following individuals and organizations contributed to the making of this book. I would like to take this brief moment to thank them and give them their deserved moment of attention in the spotlight. These are professionals in the industry who are competent and caring people.

Fred Mellis, President, Delcon Development Corporation, Los Angeles, CA
Sharon Allen, California State Legal Advisory Board, Sacramento, CA
Randall Johnson, JDC Legal Corporation, Silicon Valley, CA
Judith Reynolds, National Construction Council, Washington, DC
The American National Standards Institute, New York, NY.

Thanks and a tip of the hat to Zoe G. Foundotos, Senior Editor, for her help in the production of this book. Also I need to acknowledge Sally Glover, Editing Supervisor, and Wes Campbell, Copy Editor, for their fine work on this book. Without them, this book would have never made the long journey to your hands, your mind, and your business.

I would also like to thank all the other fine folks I have dealt with over the past years in the many departments of the McGraw-Hill publishing house who have all been outstanding individuals and uncommon literary professionals.

My best to all of you.

Chapter

1

Contract Environments

Overview

The parties to any construction contract may be of either gender and may be singular or plural—such as individuals, partnerships, or corporations. [However, traditionally, both the American Institute of Architects (AIA) and the Engineers Joint Contract Documents Committee (EJCDC), recognized nationwide as the authoritative industry standards for construction contracts, use singular, masculine pronouns as the accepted standard in text. Because this is the accepted legal industry standard, the author also uses this text format in referencing contract documents.] One of the critical factors that emphasizes the importance of contract documents and construction claims is the highly competitive nature of the construction industry today.

Because change orders (and their litigation end—construction claims) have a human cost in their normal contractual usage, change orders will (by their very nature) assign a costly mistake or oversight to some person or company. This typically leads to finger pointing and a paperwork trail to avoid expensive responsibility. Change orders bruise reputations and egos. They tend to deflate the professional image of the architect, and they invariably cost the owner money and time. By their nature and use, they create friction among the parties of the residential construction process. The major difficulty in the acceptance of this fact in contract environments is that, on balance, the greater percentage of project changes will be the result of some kind of error or deficiency. Be it the lack of or failure to secure complete contract specifications information, plain mistakes in the documents, oversights, or inconsistencies, a large portion of the changes will indicate an expensive mistake by someone.

By the time your contract has been executed, the owner has already taken full advantage of the available competition to get a low price for work that is shown and specified. The contractor is normally responsible for meeting "job conditions," but only up to a point. That point is when a *request for information* (RFI) or a *request for clarification* (RFC) becomes ambiguous as to the degree of additional work that is beyond what was anticipated and results in extra cost and time. The contractor had to assume the most competitive method of constructing the unclear details to get the job. That is, the contractor has to bid on what is shown and not price what is not shown. This practice eats up the contractor's profit in no time. This classic architect/contractor separation is clearly stated in the disclaimer of AIA Document A201, *The General Conditions of the Contract for Construction*, which states: "The Contract Documents shall not be construed to create any contractual relationship of any kind between the architect and the contractor..."

So the construction contract process always begins with the financial bond and written contractual agreement between the owner and the project's design professional(s). This agreement (contract) lays out in detail the plans and specifications of the project's construction. These two entities are the first two of the project's vested parties. There are two major professions further within the design disciplines, which are the architect and engineer. Engineers are then further broken down into structural or civil disciplines. In the construction contract environment, architectural services usually extend beyond the phase of the building's design and the production of plans and drawings. An equally important contractual task of the architect begins after the building design, cost estimates, and time schedules have been approved by the owner. The architect's knowledge of change orders and construction claims to reinforce clients' interests is a distinct business advantage. However, on any construction project of any size, changes during construction will definitely occur. And change orders are a natural part of the modern construction process.

Upon approval of the plans and specifications, the architect incorporates his or her design into the package of articles termed *contract documents*. These documents are a collection of related drawings, plans, specifications, legal notices, and contracts used by contractors to bid on a project. The drawings, plans, and specifications must be in compliance with the national building code having jurisdiction where the project is located. The architect normally also presides over the bidding, assists in selection of a bidder, and aids in establishing contracts between the owner and the contractor. The architect, in essence, becomes the agent of the owner and typically oversees the construction of the project.

During the construction process, an architect may retain consultants to supplement his or her own skills and to deliver a more complete, well-coordinated service to the owner. The architect is trained in the integration of all building systems (i.e., structural, mechanical, electrical, etc.) and has basic knowledge and design skill in each of these areas. However, on large projects he or she will rely on the design capabilities of specialists in these engineering areas. In some architect offices, the engineers are placed on retainer or under contract for services. In large offices the staff usually includes one or more of these design professionals in a variety of disciplines. The major distinction between architects and engineers run along generalist and specialist lines. Each is a true professional, certified and licensed through rigorous examinations. The engineer, however, is focused on an exhaustive study of a single technical area such as structural, civil, geotechnical and soils, or mechanical engineering. Both the design professional(s) and the contractor must have a thorough understanding of the letter and intent of construction claims in the contract environment.

Architect Responsibility

Upon building-department approval of the plans and specifications, the architect incorporates his or her design into the package of articles termed *contract documents*. As stated previously, these documents are a collection of related drawings, plans, specifications, legal notices, and contracts used by all contractors to bid on a project. Any deviation from the contract production schedule is cause for a construction claim for either the owner, architect, or builder. These are countered by a change order in the construction contract that is generated and paid for by the owner. In reality, it's usually the contractor versus the owner/architect team. This is why: The architect normally also presides over the bidding, assists in selection of a bidder, and establishes contracts between the owner and the contractor. The architect at this point is the agent of the owner. The final and equally critical point to understand here is that the architect provides overseeing of the construction of the project. This includes the interpretation of the contract documents when necessary, assisting the owner in his dealings with the contractor, ensuring that the contractor is expediting the construction in accordance with the plans and specifications, and making sure that all contractual obligations have been satisfied by the contractor.

Once assigned as his or her agent by the owner, the architect assumes what is called *agency of representation*. This means the architect speaks for the owner in decision making and has the power to authorize expenditures, or incur debt, on behalf of the owner.

Plans, Drawings, and Schedules

Architect's drawings are the main source of instruction for construction of a building and serve as the basis of the contract documents. The more specific and detailed the drawings are, the fewer the problems that develop during the construction. As a contractor, you must be sure that the plans are complete, and that you understand them, so you can plan the work effectively and estimate costs accurately. Drawings must be complete, comprehensive, and clear so the architect, owner, and builder can have a basis for contractual understanding. Architectural drawings are of two general types: those used in discussion with the customer and those used by people who construct the building. The first group consists of preliminary sketches (for discussion) and display drawings (for conveying the design to the public). The second group, often called working drawings, gives engineers and contractors plans to work from on the jobsite.

In working drawings, schedules are lists of specifications for details that recur, or repeat themselves, within a project. Equipment schedules have no connection with a planned time sequence (which is the usual meaning of the word "schedule"). Windows and doors are typically listed in specification tables on drawings, with each opening being numbered on the drawing and in the table. Because the specifications for framing and finishing may vary from one opening to another, it is more practical to table the data for each.

A full set of construction drawings and plans can include various types of drawings. To determine whether or not a set is complete, you must judge whether you have enough information to do the job. Normally, a set of drawings for new construction will include the following plans: site plan, plan view, cross sectionals, foundation plan, floor plan, elevations, and mechanicals.

- The *site plan* shows the boundaries of the property on which the building is to be built, and the position of the building on the lot. It may show hills, valleys, trees, and other features that may affect the basement or footings excavation. It may also show the nearest power lines, sewers, and water mains.
- The *plan view* shows the length and width of a project in a horizontal plane in a straight-down, bird's-eye perspective. The roof has been removed in this

view so that you can see (from above) the locations of inner walls, doors, stairs, etc. Each element is dimensioned. Its location in reference to the other components is also dimensioned. The plan view is the most widely used perspective view because it shows the project in its entirety.

- *Cross sectionals*, sometimes called cutaways, present hidden construction details and material designations (including their size and position). Cross sections are side or edge views. A section may detail a foundation, doorway, or stairway, etc. Cross sections help the plans reader understand how building components fit and work together.
- The *foundation plan* gives the dimensions of the foundation, sizes and depths of footings, location of caissons, grade beams, bond beams, plus important specifications for the load-bearing structural members. Reinforcement steel (rebar) specs are also included in the foundation plan.
- The *floor plan(s)* shows the basic layout of each floor of the building. It gives the locations of all walls, partitions, doors, windows, fixtures, and built-in features. Floor plans detail built-in features and any special interior modifications.
- *Elevations* show how the project looks when seen from the side. They show the length and height of the proposed building. Elevations show each exterior side of the building and are essential for exterior wall details of construction. They show the sides of the building oriented to the compass headings north, east, west, and south. Beyond the dimensional data, it gives the print reader a good idea of what the exterior of the building will look like. Elevations are extremely helpful in detailing features like fireplaces, alcoves, etc.
- *Mechanicals* present condensed information about the electrical, plumbing, heating, and air-conditioning systems of the building.

Construction Specifications

A full set of plans, as described above, will give a builder most of the information needed to complete a project. But, to complete it to the satisfaction of the architect, owner, and local building authority, further information is needed in written form. Such additional information is called *specifications* (which is usually simply referred to as "*specs*." Specifications usually cover the major part, if not all, of the following:

- Fire ratings for various materials

- Grades and brands of materials
- Load capacity for floors and roof
- Minimum sizes of framing members
- Spacing of joists and studs
- Gauges of electrical wire desired
- Kinds of piping permitted

Specs should be adequately detailed so there is no misunderstanding between you and the architect or the owner. If something is missing from the specifications, an extra work or change order is in the making. Always try to get the specs for every phase of a construction project. These usually include:

- Excavating	- Finished walls	- Complete roofing
- Grading	- Trim and built-ins	- Painting
- Foundations	- Rough framing	- Sheathing
- Glazing	- Tile and masonry	- Electrical system
- Plumbing	- Air conditioning	- Heating

Check plans carefully to always make sure you know which walls will carry the load. Also determine from the specs whether the existing structural supports are sound enough to carry the modifications that you're planning to add. If remodeling plans call for making openings in exterior walls for new doors and windows, plan to reinforce the structural supports where you remove portions of the wall. When cutting through roofs and floors, remember that structural members not only support the building's weight but also hold it together. You'll need a permanent support system to replace the altered members, plus an adequate temporary support system. If this is missing, change orders are apparent. You may have to arrange to move the outlets of existing plumbing and electrical systems when cutting through walls and/or floors. This should all be checked on the specs developed as part of the architect's responsibility.

Engineer Responsibility

The contractor must further understand that during the construction process, an architect may retain consultants to supplement his or her own skills and to deliver a more complete, well-coordinated service to the owner. The architect is trained in the integration of building systems (i.e., structural, mechanical, electrical, etc.) and has basic knowledge and design skill in each of these areas. However, on structural or geotechnical problems he or she will rely on the design capabilities of specialists in these engineering areas. In some architectural offices, the engineers are placed on retainer or under contract for services. In large offices, the staff usually includes one or more of these design professionals in a variety of disciplines.

The major distinction in construction law between architects and engineers runs along generalist and specialist lines. Each is a true professional, certified and licensed through examination. The engineer, however, is focused on an exhaustive study of a single technical area such as structural, civil, geotechnical and soils, or mechanical engineering. Both of these design professionals have a contractual delineation you must be aware of.

Contractor Responsibility

There are usually any number of ways a group of contractors will bid to complete a given project. The original bid will likely anticipate one reasonable sequence of construction that is apparently achievable, given the technical and time requirements. Assuming proper design workability of the architect's responsibilities, the contractor is responsible for anticipating the various components that will eventually make up the total project. Their constraints and interrelationships will be determined, and the project's parameters will be completed. Timetables are then developed and the calendar is imposed to transform the plan into a project schedule. Besides the strict technical requirements of the work itself, responsibilities of contract execution by the contractor will generally fall within these following 13 categories:

1. Duty to inquire
2. Reasonable review
3. Plan and schedule the work
4. Lay out the work
5. Supervise, direct, and install the work
6. Adequate workmanship
7. Correction of patent errors
8. Coordination of all parts of the work
9. Review, submit, and coordinate shop drawings
10. Contract payments
11. Adequate insurance

12. Adherence to safety standards
13. Warranty of clear title

Owner Responsibility

Disclaimers of owner responsibilities and warranties are frequently found in many contracts. While these must be clearly understood if they do exist, they may often be overly ambitious in their attempt to shift the owner's responsibility onto the contractor. Owner responsibility disclaimers should, therefore, be scrutinized very closely. When not barred from a specific contract, the law imposes several warranties, duties, and responsibilities on the owner, whether or not they are specifically highlighted in the contract language. They are:

1. Using sound discretion in evaluating the qualifications of low bidders
2. Maintaining and preserving the bidding system integrity
3. Funding the work
4. Providing all surveys describing physical characteristics on the site
5. Securing and paying for easements and authorizations
6. Warranting the adequacy of the plans and specifications
7. Warranting the suitability of furnished materials
8. Disclosing superior knowledge
9. Offering prompt action on clarifications (RFC) and change orders
10. Providing final interpretation of the documents
11. Cooperating
12. Assuming ultimate responsibility for the design professionals

Contract General Conditions

"Boilerplate" is a nautical term that refers to the strong, often thick material (usually steel) that forms a pressure vessel, such as a boiler. In construction contracts the general conditions are designed as an overlay that provides, among other things, a containment precedence for dispute resolution among the parties of a project by anticipating most of the areas of discussion or dispute that might arise and providing for an orderly method of resolving each case.

In contract law, the general conditions serve to "boilerplate" the vested interests of the owner and lender, by providing prearranged avenues of precedence in order to keep the production in motion during times of dispute resolution. Every possible liability is identified, critically analyzed, and prepared for. Contingency plans, insurance policies, and bonds should be obtained as collateral for every dollar of your projected profit in the project. Every clause in every one of your contract documents should be completely understood and agreed to so that its contents focus squarely on protecting the production and timely completion of your contracted project schedule.

In addition to the working relationship provisions that they contain, the contract's general conditions are not limited to the production portions alone. Typically, all things that are bound into the contract's general conditions are considered part of the primary contract document. These may include the bid bonds, agreement contracts, performance and payment bonds, and conditions of the contract. By law, construction contracts encompass general conditions, plans, relative contract documents, specifications, and technical specifications addenda. Generally, construction contracts can be broken down into three categories, though they are not necessarily arranged in the same order on each job. These categories are:

Part I: Bidding and Contractual Documents

- Notice of Invitation to Bid
- Instructions to Bidders
- Proposal Bid Forms
- Bid Sheets
- Contractor Certificates (Licensing and Surety Bonds)
- List of Subcontractors
- Bid Bonds
- Non Collusion Forms
- Agreement Contract
- Performance Bonds
- Payment Bonds

Part II: Conditions of the Contract

- General Conditions
- Supplementary General Conditions

Part III: Specifications and Specifications Addenda

Architect or design professional therein provides technical sections covering the various parts of the project.

Each of these items has further details, but an in-depth examination of each is beyond the scope of the contractor's responsibility and falls instead in the lap of the owner and designer. The architect or design professional is responsible for supplying technical specifications as well as the design plans. These contract specifications and their addenda are added to the contract's general conditions to explain in detail the various responsibilities of the architect or design professional regarding the review and approval of shop drawings, as well as the parameters of design to production liabilities. Contractors need to be aware of all of this information to prevent unclear or incomplete details specifications from causing delays in their schedules.

For contractors, reviewing the general conditions, specifications, and specifications addenda is a critical step to ensuring contingent activity workarounds and that long-lead items do not burn up valuable float time, thereby placing your completion date in jeopardy. Prime contractors also need to function within the general-conditions responsibility as they subcontract for each contract specification—per plans and specifications.

Contract Specifications

Contract specifications are not limited to the technical portions of the contract alone. Typically, everything that is contained within the specifications document is referred to as "the specifications." These may include the notice of invitation to bid, the bidding documents, bid bonds, agreement contracts, performance and payment bonds, noncollusion affidavits (where applicable), conditions of the contract, technical provisions of the contract, and provisions of the contract project schedule. If you encounter plans calling for a book of standard specifications by reference, this may signal a specifications writer who was not proficient at detailing specifications. Use caution here; this is an area of claims generation. Every project lender, contractor, and owner who has spent a few years in the trades has encountered specifications that were questionable at best. (I'm being generous here because this is an understatement.) Construction management professionals will tell you that not all specifications writers are knowledgeable in all the subjects they are writing specifications for. Sometimes they're just flat wrong.

If an error in the specifications is not caught by the responsible contractor before the work is done, the production schedule will be pinched by the dispute resolution necessary to unscrew the situation. It often happens and is one of the things you need to be aware of that can stall an otherwise viable project schedule. A comparison of the finish schedule with the specification index will expose potential duplications and/or omissions that will invariably lead to change orders. It will indicate whether the design process was completed in an organized and sequential manner, or put together by different people who never worked together. Any discrepancies discovered in this review will be fairly obvious. Job logic here is another place to look for confusion.

Comparison of the specifications to the finish schedule includes analysis of the schedule's headings against the categories included in the technical specifications. You're looking to confirm that each item is noted and accommodated in the technical specifications, and that each item is included only once. Should any discrepancy or duplication be discovered, the architect or design professional must be immediately notified of the error. Some design professionals have tried to limit the extent of their liabilities by including exculpatory clauses in the specifications addenda, which usually read something like this:

"Plans and specifications provided are complimentary. The contractor is responsible to provide all work shown, whether or not adequately described in these specifications, and all work described in the specifications, whether or not specifically indicated on the plans, to code compliance and current accepted workmanship standards." If you see something like this in the specifications addenda, cringe. You're looking at what construction management professionals call a "weasel clause." The design professional was not an expert in that technical specification and/or didn't research it thoroughly enough. And if it wasn't worked out in the design process, the only other place it can be worked out is in production, blowing your project schedule and costing lots of money. Unless, of course, you're smart enough to be looking for these problems ahead of time. I equate weasel clauses to dancing. Every time I see one in the specifications addenda, I know it's time to go cha-cha with the architect.

Specifications and their addenda are needed in the contract documents to provide comprehensive technical explanations of project systems and their function. They must be complete for the production process to flow smoothly. If there is an absence of clear definition and delineation of these responsibilities, change orders and claims are the outcome. As a member of the project team, the contractor needs to be aware that the courts have held that the basic design responsibilities listed below are considered the *architect's* contractual obligation, whether expressed or implied, within the specifications or specifications addenda:

- Code compliance
- Technical accuracy of all documents

- Producing all plans and specifications
- Specific design (not to be substituted for design criteria)
- Interpretation of the documents
- Standard workability of design, as per specs
- Shop drawings submittal review and approval
- Prompt and timely response in review approvals as to not delay work
- Evaluation of the work
- Due diligence in judgment and professional expertise

In addition, the court has held that the first responsibility of the design professional is to indicate clearly and completely all work in sufficient detail, on the plans and specifications, as to adequately describe the parameters of each of the specifications. So this ultimate responsibility to comprehensively and adequately describe each construction specification lies with the design professional who generates the plans and specifications, regardless of any exculpatory clauses. Disallowing such clauses, the court has held that design professionals are ultimately responsible for technical specifications and their related responsibility identification. Requests for information (RFI) are sent from the job site back to the designer for anything overlooked or incomplete in the specifications, by the field contractors. When too many RFIs are sent back to the architect, it signals any experienced owner/developer that it's time to hire another architect on the next project.

Remember, specifications are not limited to the technical portions alone. Typically, everything that is contained in the specifications document is referred to as the specifications. These may include the notice of invitation to bid, the bidding documents, bid bonds, agreement contracts, performance and payment bonds, noncollusion affidavits (where applicable), conditions of the contract, and technical provisions of the contract. Contract documents, by law, encompass all plans, relative documents, and specifications.

Progress Schedule Regimentation

At this stage of the contract project schedule planning, the agenda should be to compile a complete list showing all milestone events and their respective dates. These should include all meetings, prime and subcontractors progress payment schedules, progress inspections and testing requirements, equipment delivery dates, long-lead items delivery dates, submittals, and project phases completion dates.

As contracts are awarded, all the subcontractors should be asked to submit their own activity schedules to see how they fit into the overall project schedule, and to submit a list of proposed shop drawing submittals and delivery dates of the equipment after receipt of approved drawings. When these schedules are received, the prime contractor must review them to ascertain that the equipment delivery dates fit into the overall time frame of the job. If they do, then they should be incorporated into the project schedule and distributed to all subcontractors. Subcontractors should be given the complete project schedule containing their portions of work, and they should be asked to comment on the length of time and sequences allotted to them. With their input initially, and regularly updated status reports, the schedule will be more accurate and less likely to generate change orders and claims.

Once the prime contractor has identified the activities' work items at the subcontractor microcosm level, the activities listed on the primary report file (sort by activities) are then broken out from the sort-by-activities time line and transposed onto the accelerated S-chart time line. This, in effect, shifts the production network into double-time. Superimposing the network time line S-chart on the sort-by-activities bar chart opens up details subtasks and work items within the activities and thus, windows of float. The duration of events opening and closing begins to build a graphic fabric that should reflect the job logic. Bottlenecks appear and workarounds are developed. The overlaid events and milestones should become workable, and the job logic of the network schedule should flow in an organized manner. If not, the project will be riddled with change orders and claims.

Chapter 2
Contract Conditions Conflicts

Claims Consciousness

In contractual obligations and construction management of project operations, claims consciousness is the knowledge that even seemingly insignificant problems can and will escalate through neglect or mismanagement to sizable liquidated damages. Each of the players in the project has investment of time and capital in the project and each has something to lose if the project is not completed according to contract. Delay of the project schedule alone can bankrupt a small contractor's profit margin, let alone fixing the bottleneck problem out of pocket. And as the complexity of today's contracts and projects increases, so too does the potential of problems leading to change orders and construction claims for damages.

Claims consciousness, at its primary level, means educating oneself and understanding where change orders and claims originate. Steps can then be taken to analyze and alleviate their impact before they grow into financial alligators. Claims consciousness, at its secondary level, is using this investigative step for preparation and prevention of situations leading to change orders and construction claims. This is done by careful analysis of errors and omissions, schedule delays, schedule accelerations, interferences, and latent conditions.

Claims consciousness, at its final level, means winning the claims litigation with favorable judgment and settlement. This is achieved by mastery of the first two levels. Additionally, from such a contractual position of strength, the disputed parties in construction claims clearly see the advantage in a quick settlement as opposed to a protracted and costly litigation.

Modern contractors have now come to know their contracts better than the designers who produced them. Contractors also control their contract administration more efficiently and effectively. What the architect would have previously just clarified through an RFC in follow-up is now a change order for which the contractor will get paid. Many contractors have become very good at getting what they are after in claims management. When necessary, through contract dispute resolution, they recover damages through the construction claims process that culminates with an ending judgment through the court or arbitration.

In the last decade, these construction claims have become hot news items because the dollar amounts won in many of the awards were very large. The awarded claims became testament to the fact that contractors clearly did not have to take abuse any longer in the administration of a construction project. The concept of being paid for a time delay was much more attractive monetarily than the historic "granted time extension" tactic of the past. Contractors, once they began to see the winning side of construction claims, became very interested in developing the techniques outlined in this book. Some prime contractors even began to look upon the whole idea of change orders as a profit center in itself. The tactic of "low-balling" (contractors intentionally submitting unrealistically low bids to get project awards while all the while plotting to make up those dollars through claims and legal loopholes) came into being and soon gained national notoriety. This came at a time when designers were still refusing to admit that the root of the problem was change orders and claims resulting from flaws in their own contract documents.

The American Arbitration Association (AAA) began accepting construction claims cases for arbitration in 1966, but didn't begin to maintain comprehensive records until 1973. From that point, the number of construction claims arbitration cases submitted per year nationally doubled in the decade that followed, and continues to increase at the same rate. Even if a construction company is not particularly interested in pursuing an aggressive claims posture, a responsiveness to the system has now become absolutely necessary to at least maintain an ability to accurately assess day-to-day operating problems as they occur. A working knowledge of contract structures and dispute resolution options has now become the most fundamental prerequisite for responsible contracting management. The most advantageous positions for defense, protection of rights, and the eventual presentation of a claim (if one becomes necessary) must be pursued.

Claims consciousness begins with an awareness that even the seemingly most insignificant problem on a construction project may, through mishandling or neglect, work itself up to devastatingly expensive proportions. This awareness is combined with understanding that increased complexity increases the probability that serious problems will have to be dealt with on nearly every construction project of any size. Therefore, first and foremost, contractors must become familiar with all literal and implied meanings of their contracts. Secondly, they must know their contract environments. By this I mean the state of affairs in the modern construction industry; operations and perspectives of designers, subcontractors, and owners; and the current precedents set by the court in similar construction claims cases.

So claims consciousness is a frame of mind. The contractor learns to view each problem as it occurs as either a potential serious deadlock or a potential change order. Claims consciousness then becomes a matter of prevention or occurrence, whichever is the objective. Many problems in construction fall within recurring categories, and these can be deduced and predicted rather accurately. Errors and omissions, project schedule accelerations, schedule delays, interferences, and latent conditions will make the top of most lists.

Claims Generation

Change orders that are rejected by the owner often get returned cut up so badly that they can hardly be recognized, or they are flatly rejected. Handling these rejections successfully depends on completely (and correctly) understanding the professional and personal motivations of your opponent. This must be first accomplished to move further in the correct direction of successful resolution. Next comes determining those actions that you can take to satisfy the owner's motivations you have identified. Do not overlook the personal motivations in your assessment.

Weigh each possible action the owner may take against the actual benefit to be realized if it accomplishes the intended objective. Build a strategy with your assessments of each probable action. Be prepared to move forward quickly in the next step in the contractual dispute resolution process. Know the next step in advance and be planning for it. Demonstrate your commitment and resolve in getting paid for the change order(s). Be clear about your complete intentions to follow through, and then follow through when it becomes necessary.

An owner or its agent will typically respond in a tough and unyielding manner due to the loss of money involved. This is typical business procedure and should not be misinterpreted as meaning the owner is being unreasonable or unprofessional. Modern construction contracts, with all the associated performance upon which they are contingent, are extremely complex. This holds true for not only what is stated explicitly in the contract documents, but also for what is implied.

The first question to consider upon receipt of a change order rejection is simply whether or not the rejection was made in good or bad faith. Check carefully to see if there is any contractual basis upon which the rejection is based. If so, this is deemed a "good faith" rejection and will be looked upon by the court as having originated from at least an attempt on the owner's part to responsibly evaluate the contract before issuing the response. The contractor's response can likewise be based upon a clear application of those portions of the contract that support the change order claim. Upon analysis it may become apparent that the cost of fighting an unwarranted action will meet or exceed the amount of settling the rejection.

Construction dispute resolution differences that have gone as far as they can without being resolved are then normally subjected to one of two different dispute resolution options: arbitration or litigation. There are important differences between them, and your awareness of the advantages and disadvantages of each will allow you to determine which you will prefer for the specific case you're considering. Arbitration will not be a dispute resolution option unless there is a specific agreement between the parties to arbitrate. This agreement can be in the form of an existing clause in the contract under which you're working or a separate agreement between the disputing parties drawn up at the time of the dispute resolution. On balance, the second situation is very difficult to accomplish if you are the one to determine that the case lends itself best to arbitration. The owner will see the disadvantage from his or her perspective and a subsequent agreement to arbitrate will then become very unlikely.

Production Schedule Change Orders

In dealing with change orders in the production schedule, remember, contract law does not allow the prime contractor or subcontractors to make any changes to an existing contract. An addendum in specified work requires the contractor to write up a contract change order and then seek written approval from the owner. The owner is not obligated to sign a change order, but the owner's approval is absolutely necessary to avoid litigation problems. The owner cannot be forced to pay for work that was not authorized by the owner. Contract law specifies that "all changes to the original contract must be in writing and signed by both the property owner and contractor." This is relative to all those claims generated from project schedule problems which are known as *assigns and deducts*.

There are six fundamentals of production scheduling that form the basis for presentable evidence in a court of law for all construction project schedules and will, in turn, establish the basis of good recordkeeping, clear accountability, and effective legal presentation. The first of these is that the project schedule must be the same one that was actually used to build the project. Even if a schedule that is substantially different from the one under consideration had been formally submitted and approved by the owner and/or the design professional, the schedule that was actually used and depended upon by the various trades will generally be considered to be the contract document.

Second, the project schedule must be periodically revised. It is generally recognized that project scheduling, however exact or inexact, is nothing more than a plan to construct what can reasonably be construed from the contract documents, considering that there may be several ways to build the same project. Although contingencies may be included to allow for imprecisely defined variables, there probably never has been a schedule that anticipated every problem, coordinated every piece of work, and required no modifications to make it work. Changes to, or corrections of, the plan are inevitable. The schedule, therefore, must be updated periodically to maintain a current and accurate representation of reality.

Third, the periodic updates of the project schedule must show all positive and negative influences by all parties. None must be singled out; none must be absent for convenience. If all updates, for example, only indicate delays caused by the owner and fail to delineate other known problems caused by other parties, it won't be very difficult for you to demonstrate that bias has been built into the document. Failing to recognize all significant events will bring into question the validity of the entire project schedule defense presentation.

Fourth, the project schedule must include realistic construction logic and activity durations to demonstrate the professional ability of the planning and scheduling team. Job logic sequences that are out of order or the lack of consideration of critical variables (such as indicating the installation of HVAC units in an area with inadequate access, to occur after the partition walls are constructed) will only demonstrate that the constraints in the project schedule were unrealistic to begin with.

Fifth, the project schedule must fairly represent the actual method planned to build the project. If, for example, the project schedule had been prepared primarily to cater to progress payments, or was otherwise unrelated to actual construction of the project, it would become clear that, as a tool for managing the project, its value was marginal at best.

Finally, the project schedule updates and analyses must be realistic and in perspective. An overly aggressive computation of damages might hurt the validity of the entire analysis. Direct cause-effect relationships must be shown.

In addition to the foregoing, the project schedule must contain all notes and references that are correlated with the more detailed, chronological project correspondence. Have all field notes, claims, commitments, deadlines, and promises detailed in the schedule documents. Include the dates and the names of the individuals in those organizations supplying information. Pin it down and document everything. Litigation is warfare.

Guaranteed Maximum Price (GMP)

In this type of construction contract, the contractor is responsible for completing the project for a total sum equal to or less than the guaranteed maximum contract price. What is more, the contractor may be required to sign off to the initial GMP with only 60% to 80% of the drawings complete. The original quoted GMP may be subject to adjustment based upon changes in scope, but even that knowledge cannot be considered as a blank check. The contractor must continually exercise extreme caution in the application of his or her superior knowledge in the construction process. In this type of contract, the prime contractor now contracts directly with, and is legally responsible for, each respective subtrade.

Critical Path Management (CPM)

Professional computerized project scheduling is built on the fundamentals of the critical path method, or CPM as it is known in the trade. Critical path method refers to identifying bottlenecks in the production process and then building a production time line through them, using the total duration of those activities which will take the longest to complete as the basic time scale. Critical path management is the accepted method for managing those time line paths efficiently by manipulating time management of the interlinkage of activities durations and contingency workarounds. It is a scheduling system that allows the project contractor to achieve improved time control over a project's production phases. Additionally, an audit trail and cost tracking can be incorporated into the system to provide a reasonably accurate estimate of time frames required for those or similar activities in future projects.

CPM typically works best in straight-line, time-scaled productions in construction where the building time lines can be estimated with a fair amount of certainty. These network scheduling systems are essentially a combination of CPM and old-style hand-drawn bar charts called S-charts. Critical path management is the most accurate computerized system of network scheduling yet developed. It allows owners and contractors to achieve control in critical areas of project scheduling. CPM has proven itself by adding more bottom line to owners' investments. In larger projects, such as multiresidential and commercial, CPM is mandated by contract.

All production activities will affect each other in either direct or indirect linkage. CPM fundamentals require that a critical activity must be finished before the succeeding critical activity can start. Noncritical activities and those of similar trades, such as roughing in the plumbing and electrical work, can proceed simultaneously. One of the objectives in critical path management is to find the order (critical path) in which the phases of critical activities must be completed. Those designated as critical activities begin by being assigned to the production activities that will take the longest to complete and are crucial to the production time line. By tracking these activities, attention is automatically drawn also to those areas where it is most essential to avoid production delay. These activities are sources of contract conditions conflict. When bottlenecks occur in the schedule, those activities and their subactivities on the critical path must be handled first, while those delays off the critical path can be addressed secondarily. This is especially true if a particular activity has zero or negative float. Delays along the critical path begin affecting total project duration. Critical paths may change as production bottlenecks are broken, critical subactivities are changed, or when new problems surface.

CPM is currently the only construction scheduling system that covers all phases of a project and allows the contractor to manipulate time frames during and around the activities that will take the longest time to complete, then select the best ways to expedite or work around the phases of activities for contingency planning. The underlying logic in critical path management is that contingency planning for every activity and every project phase reduces risk.

When this time-scale interlinking of production activities is combined with a software tracking system, the contractor can then provide computerized audit trails and cost-tracking reports, as well as trends analysis, which is next quarter's forecast of the current numbers logical outcome. These are crucial functions for protecting your money interests in the project.

Cost-Loaded CPM

Cost-loaded CPM comes into use with unit-price contracts. These contracts use prices that are based upon an itemized breakout list of all work activities involved in the contract. Each is assigned a unit price. Using this method, the final line-item price on each activity item is computed by multiplying the unit price times the item quantities. The unit price is then used as the controlling price for each item. Cost-loaded CPM is built from the unit-price bid sheets supplied in the original contract by the activities subcontractors. This is used by the contractor as a schedule of values, and must not be confused with the lump-sum method. In a lump-sum project, the contractors contract to do all the work defined in the plans and specifications for a fixed price stated in the contract.

Any quantity variations that are not added by the project team as change orders are not used in factoring. To cover these possible variations, a schedule of values is also requested from the contractors in their bids. A schedule of values is basically a unit-price bid with some notable differences. For example, item prices are shown for ease in making monthly progress payments only; they cannot be used to price change orders and have no contractual obligations. Cost-loaded CPM is used to advance phased scheduling on progressive milestones.

A common problem that occurs when owners request cost-loaded CPM is their assumption that cost-loaded CPM will always result in easier progress payments management. This is true of lump-sum projects, but the situation is reversed in unit-price contract projects because a cost-loaded network does not synchronize with the pay line items on unit-price contracts. Therefore, in a lump-sum contract with cost-loaded CPM, the unit-price breakout should be submitted after the contract approval but before the first progress payment. In a lump-sum contract without cost-loaded CPM, a schedule of values should be submitted after the contract approval but before the commencement of work.

Lump-Sum Contracts

In a lump-sum contract, the contractor agrees to build the complete project as per plans and specifications. Anything necessary to accomplish the completion of the project must be considered part of the contract, even if not specifically stated in those plans and specifications. Conversely, in a unit-price contract the contractor may still be required to construct a complete functional project, but if any variation occurs in the quantities of any of the separate bid items listed, the contractor is entitled to an amount of money equal to the unit price of the bid item, multiplied by the actual quantity that was supplied or installed of that item. As an example, typically most engineering projects are based on a unit-price contract where the actual unit prices stated in the original bid will be held as the factor for all progress payment disbursement amounts. Because a unit-price contract does not have a fixed ceiling price, the final cost to the owner will be determined by the quantities actually completed. To adequately protect profit margin, the contractor must be careful to determine that the field quantities are precise. Conversely, in a lump-sum contract project, if an error is made within the first month of work it can be compensated for in a later progress payment. The contractor cannot receive more money than the stated fixed price, even if some of the final quantities varied from the anticipated amounts.

Daily Field Reports

Daily field reports are the fundamental documents that record actual job progress, together with all conditions that affect the work. Because of that, they are usually pivotal in contract conditions conflicts by establishing cause and effect of the change order. Daily field reports provide the progress reporting basis for the actual results of the schedule as compared to projected progress, this "actual versus projected" difference being most often the primary initial source of claims. They begin an organization's standardization of reporting at the source of causes and effects—the point of production. If actually prepared daily, they are generally considered by the court to be the best sources of job documentation information supporting or denying construction claims. This is because they are supposed to be prepared immediately as the information is being generated, with no appreciable time lapses. The inclusions and descriptions are fresh in everyone's mind. This facilitates complete details of the work. In addition, they are prepared by those with authority and responsibility for the work. They have an interest in the accuracy of the information and have a significant incentive to maintain the report's accuracy and completeness. These are the people who actually witnessed the work and are recognized to be the most qualified to describe the reported facts. Because they will be the most detailed, accurate, and complete records of all job site events, the daily field reports will be used by the court to resolve any dispute about what actually happened on the job. They are so valuable to the contractor in prospecting for change orders because of the wealth of information recorded in summary form. This data includes:

- Work and activity descriptions, separated as to the physical location and extent
- Labor force, broken down by subcontractor, locations, and major activities
- Equipment used and stored by the various activity subcontractors
- Site administrative staff and facilities
- Weather conditions and temperatures at key times of each work day
- Change order work accomplished, with relevant details
- Photographs taken during the day
- Visitors to the jobsite
- Meetings, discussions, commitments, and conversations

Never use the back of a field report to continue information because it will be overlooked when faxing, during a review, or photocopying. Because it is impossible to tell in advance what will become the most important information, all categories must be accurately maintained. It is up to you to organize the reports in your own way. They must be checked often enough to ensure that all required information is being properly recorded.

What is more, the requirement that the reports be completed *daily* cannot be stressed enough. If they're allowed to be lumped together for completion at the end of the week, the accuracy of the information quickly degenerates. The completeness of the data also is dramatically reduced. When on site, you should visibly refer to the reports often, during conversations, issue research, and so on. Let your field personnel know that the information is depended on and used, and that their input is extremely important. Even if you don't really need to at the moment, come up with an excuse for looking at the reports. The minute that the field personnel begin to get the feeling that no one's looking at the information, the data will become very thin indeed.

Have the reports printed on two-page carbon or NCR paper. The original is to be sent to the central office, with the copy retained in the field office. Originals should be faxed over at regular daily times, with the original sent or picked up for recordkeeping. Requiring daily deliveries or mailings of field reports further ensures they will be done daily, in a timely manner. The daily field report is typically filled out by the site superintendent and signed by the project manager if that is an on-site position. In the absence of a site superintendent on smaller projects, the report must be filled out by the responsible project manager on a daily basis. Field reports need to include information listed in these pertinent areas:

- **Title Box**. Indicate project name, company job number, project location, and the name of the individual completing the report.
- **Date/Page**. Account for each page of the total report. Use as many report forms as necessary for a complete record of the day's events.
- **Weather**. Include a short comment on the typical condition of the day (rainy, cloudy, sunny, etc.). Record the temperature at the beginning and the end of each day.
- **Visitors**. List all visitors to the job site that day who are not part of the regular work force. Include a brief remark as to the reason for the visit. Note any special situations deserving immediate attention. Include references to appropriate documentation.
- **Work Force**. Indicate the number of each type of employment classification for each type of labor involved.
- **Activity Performed**. Directly adjacent to the respective labor force information, include a short description of the type of work performed.
- **Equipment**. List all major pieces of equipment on the site, whether or not they arrived on that day. Include brief descriptions of equipment work performed.
- **Items Received/Sent**. List all items received or sent that day only.
- **Location**. Provide sufficient references to allow accurate location of the work. For example, "Continue foundation wall forming along column line A, between lines 4 and 6."
- **Change Order Number**. If the work described under "Location" applies to a change or a proposed change, insert the number of the change order or change estimate number in the column provided.
- **Problems/Comments**. Delays of any kind noted here. What, who, where, and when. Time in and out of any production hiatus.

Summary Reports Conflicts

Summary reports are totals of all the individual estimated, projected, and actual information current in the project as of each report's run date, and they must be organized and distributed during project operations. A computerized project schedule has reports that are interlinked with one another and provide summary reports that must be sorted out to the appropriate project team members, owner, prime contractor, subcontractors, and procurement suppliers by the type of information needed by that specific reader.

The large menu of output summary reports is another big advantage of a computerized CPM schedule. However, the information is only useful if you receive it. Insisting on updated summary reports allows the various members of the project team to receive the same summary reports of progress, pending construction, or future scheduling that might lead to production bottlenecks or change orders. Most computerized summary sorts will yield a sort menu of summary reports as follows:

1. Sort by activities
2. Sort by *i-j* numbers
3. Sort by early starts
4. Sort by job logic
5. Sort by project milestone
6. Bar chart printout
7. Human resource leveling
8. Key-milestone-date sort
9. Critical item sort
10. Sort by events
11. Free float per activity with total float summary
12. Critical-path sort
13. Critical-equipment sort
14. Sort by total float/late start
15. Limited periodic look-ahead sorts
16. Schedule of anticipated earnings
17. Cost by activity number

Most contractors find the sorts by total float and by milestone most valuable for their needs. The total-float sort starts with the low-float (most critical) work activities listed first for immediate attention. The less critical high-float items show up later on the list. By using periodic look-ahead sorts, one can also home in on specific time periods. A 30-, 60-, or 90-day look-ahead sort will list only those critical items that will occur in the next 30, 60, or 90 days. These sorts are extremely useful for the contractor in prospecting for potential change orders.

Other members of the project team need other types of data sorts to make their work more productive. For example, materials-control people find the critical item sort more convenient in tracking their required delivery dates than extracting those dates from a milestone chart. As the revised delivery data and actual progress are fed into the computer, revised printouts quickly reflect the delivery changes and their effect on the field schedule. Field people usually find that the key-milestone-date sort better suits their needs. That section of the project schedule is the basic document that the prime contractor uses to make the detailed weekly work schedules in the field.

The rapid turnaround of data sorts by the computer also allows the contractor to perform "what-if" exercises with the logic diagram. When scheduling problems arise that appear headed to change orders, the contractor can try alternative solutions for a workaround by reshifting elapsed times for problem activities. This generates new early and late start dates that can be shifted to improve the critical path. The computer calculates a new critical path in a matter of seconds, with immediate access to the new output data sorts on the computer monitor, and the contractor has its change order and workaround accepted.

If it is desired to determine all items that can be started as of any given date, reference to this sort will show under the column headed "Event Start," subhead "Early," the earliest date that any activity can be begun. On-the-schedule reports show that the effect of using total float for a given activity can be shown. Activities that are in series are consecutive operations, a fact that can be noted on both the project operations printouts and the detailed project schedule plan. By altering the events of other activities in the network, days of slack can be realized for the activities not on the critical path schedule. Zero free float in activities red-flags the contractor that this activity is heading into negative float, which in turn robs total float from the entire project schedule.

Recycling the Schedule

Monthly, the project schedule should be recycled. This means to put it to critical analysis and cook it. Float, elapsed-time estimates, summary sorts, and critical paths and trends analysis forecasting take first priority on the agenda. Critical path activities that met scheduled events need to be analyzed for the characteristics of success that can be incorporated into future project schedules. Critical path activities running in negative float need to be straightened out.

A computerized system simplifies recycling the schedule. Besides the traditional quarterly reworking of the schedule, recycling becomes necessary whenever schedule deviations accumulate to a point at which some of the intermediate milestones are in jeopardy. Recycling involves revising any target dates that may have slipped beyond repair, perhaps because a significant change in scope has occurred. Exercising some "what-if" options allows the savvy contractor to make the best revisions.

Having activities with negative float is a sign that the schedule no longer is achievable in its present form. Recycling the schedule should not be confused with the monthly progress evaluation. Recycling is necessary at the end of the month if your short-term milestones are not being met. The recycling procedure critically analyzes how to get the project back on track without extending the completion date. When recycling the schedule, keep juggling and rechecking the resulting logic diagram for the best job logic.

The best place to start recycling is with the network's job logic. Remember, the logic diagram is the central command center of the project schedule. If the program logic doesn't wash there, that's the beginning of the end. Nothing else, from the logic diagram outwards in the network will be running in sync. Fix this crucial component of the schedule first. Then move into the next stage of schedule recycling, which is operations analysis.

The simplest scheduling recycling that will avoid contract conflicts is concerned with *time* use. Thus, you might allocate 10 hours each week in project cost-tracking and audit trail, and 20 hours to sorts and their communication with appropriate project team members. You might dedicate the last day of the week to end-of-month sorts and the summary reports, and an early hour every Monday to preweek schedule management planning.

A more complex scheduling recycling system compartmentalizes by *activity function*. Calculating the percentage of each activity's progress towards completion is the responsibility of that activity's subcontractor, but the summary report of that progress is the prime contractor's responsibility. These more complex systems tend to be (or become) hierarchical. They simplify delegation of each activity's responsibility and according sorts but de-emphasize total project objectives and are slow to adapt to changes during schedule recycling. In planning the schedule activities and operations analysis, it is important to reflect the objectives of the owner and the project team. The degree of project team control versus prime contractor and subcontractor empowerment, and the overall project matrix network, will determine in significant part whether you will reach the project's phases milestones on time or be headed for change orders. This is a trade-off or averaging tool the contractor uses between time scaled activity events and activity control, to accelerate or constrain the activity.

Chapter 3

Clarifying Contract Change Orders

Who's Who

The construction contract process begins with the design professionals. The two major professions within the design disciplines are the architect and engineer. Architecture is, in essence, the pursuit of the art and science of habitat for man in his environment. However, architectural services usually extend beyond the phase of the building's design and the production of plans and drawings. An equally important task of the architect begins after the building design, cost estimates, and time frames for the project schedule have been approved by the owner. But the contractual responsibilities are not finished there. Upon approval of the plans and specifications, the architect incorporates his or her design into the package of articles termed *contract documents*. These documents are a collection of related drawings, plans, specifications, legal notices and contracts used by contractors to bid on a project. The drawings, plans and specifications must be in compliance with the national building code having jurisdiction where the project is located. The architect normally also presides over the bidding, assists in selection of a bidder, and aids in establishing contracts between the owner and the contractor. The architect, in essence, becomes the agent of the owner. In legalese, this is called *agency of representation*.

The final and equally critical service the architect provides is overseeing the construction of the project. This includes the interpretation of the contract documents when necessary, assisting the contractor in his dealings with the owner, ensuring that all contracts are being expedited in accordance with the plans and specifications, and that all contractual obligations have been satisfied by all parties to the contract.

During the construction process, an architect may retain consultants to supplement his or her own skills and to deliver a more complete, well-coordinated service to his client. The architect is trained in the integration of all building systems (i.e., structural, mechanical, electrical, etc.) and has basic knowledge and design skill in each of these areas. However, on large projects he or she will rely on the design capabilities of specialists in these engineering areas. In some architect offices, the engineers are placed on retainer or under contract for services. In large offices the staff usually includes one or more of these design professionals in a variety of disciplines.

The major distinction between architects and engineers runs along generalist and specialist lines. Each is a true professional, certified and licensed through rigorous examinations. The engineer, however, is focused on an exhaustive study of a single technical area such as structural, civil, geotechnical and soils, or mechanical engineering. Both these design professionals must have a thorough understanding of their liability in errors and omissions that lead to construction claims.

Change Order Compensation

To identify clearly the contract elements that will eventually sum up the maximum possible compensation allowed by a contract change order, visualize as completely as possible the two scenarios being constructed. First is the work as originally planned and as it actually would have been completed without the subject interruption? Second is the work all over again, this time with all the impacts of additional scope, rescheduling, payment disruptions, and so on?

As the changed sequence is considered, catalog all extra cost items as they apply to the three costs of every change: direct costs (labor, materials, supervision, etc.—the *hard costs*); indirect costs (home office, opportunity costs, lost profit—the *soft costs*); and consequential costs (interference, disruption, rescheduling). These three cost components of every change order must be detailed and broken out. Their applicability to a particular change are matters of degree. Their appropriate inclusion in the respective change proposal is a matter of business judgment, applied within the context of the owner-contractor relationship.

There are typically ten reasons for change orders, what are called in the trades "adds" and "deducts." These ten reasons are: design errors (contradictions, discrepancies, impossibilities, and inconsistencies); changes in market conditions (specified product becomes unavailable, new products become available offering better price or performance, new information becomes available affecting the choice of specified materials); change in the owner's requirements (scope of work); uncovering of undisclosed existing conditions; uncovering of unknown existing (latent) conditions (unexpected soil variations, conditions uncovered during alterations to an existing structure); suggestions to initiate better, faster, or more economical construction; change in designer preference; discrepancies in the contract documents have described situations contradicting the intent of the project; change in external requirements (building codes, zoning, or other changes); and final coordination with N.I.C. (not in contract) equipment.

Most change orders will apply to one of three distinct categories: owner-acknowledged (initiated or approved by); constructive; or consequential. The applicable category will determine the best formal process, informal and political approaches, and probably for resolution and payment without conflict. Let's look at all three of these categories in greater depth.

- Owner-acknowledged change orders: This is the category that most commonly comes to mind when the subject of change orders is considered. It is the only change type that will become incorporated into the owner's budget and will ultimately present the least conflict. The primary objective of proper change-order management is to convert the change orders of the other classification into this type. This is necessary because this is the only category in which the owner will write a check without being ordered to do so by the court.
- Constructive changes: A constructive change order is one that occurs when the owner or his or her authorized representative acts in such a way that causes a contractor to perform additional work. These kinds of actions or inactions may take the form of verbal and written directives (outside any specific change order procedure), but may also include any act or omission (failure to act) that has the ultimate effect of changing the work from that specified in the contract. There are seven typical categories of constructive change orders: changes in method of performance, misinterpretation of specifications, defective specifications, overinspection, rejection of "or equal" submissions, defective specifications, and defective owner-furnished property or equipment.

  There is usually no formal recognition initiated by the owner that a contract change does in fact exist. Since no change order is accordingly issued, the contractor must be quick to recognize the action or inaction for what it is. The conduct that results in the change must be identified and documented as early as possible, if any hope to secure additional compensation can be preserved. The ultimate objective here becomes the conversion of the constructive claim into an owner-acknowledged one.

- Consequential changes: A consequential change is similar to a constructive change. It differs in that it is additional work that becomes necessary or additional costs incurred as a direct result of some usually more obvious change. It is the result of a domino relationship of cause and effect. The cost components of a consequential change may include any or all of the components of any of the other change order types. These would include direct cost of additional work, interference costs (rescheduling, rework, etc.), and impact costs (delay, extended overhead, opportunity cost, etc.).

Concessions made by the contractor in these change-order categories must be based on the reality that the value of materials and labor is greater before they are rendered than after. Therefore, some basic rules of concessions:

- Give yourself room to make some concessions. Have a good reason for starting where you do, and be able to substantiate it. Start high.
- Make your opponents work hard for everything they get. People don't appreciate something for nothing. The harder they work, the more they'll appreciate it.
- Conserve your concessions. Later is better than now. Make your opponent work for every concession you are willing to make.
- Avoid negotiating away or swapping concessions. If your opponent says, "Let's split the difference," consider saying, "I can't afford to."
- Train yourself to get something for every concession.
- Don't be afraid to say no.
- Don't lose track of your concessions. Keep a list.
- Don't be afraid to back away from previous concessions. The agreement that counts is the one at the bottom line.

Change Order Finalized

By the time the changed work is begun at the jobsite, accurate documentation becomes absolutely necessary. This is true whether or not the change order has been finalized at that point. If the change order has been finalized, the first consideration is to maintain a comparison of actual costs versus estimated costs. If costs begin to get out of control, it is imperative to be aware of the situation at the earliest possible moment. Just because a change order has been executed, it does not necessarily mean you are stuck with the existing bottom line. Any number of effects may have been introduced into the actual work that can justify your reopening of the issue. These might include cost of overruns due to conditions unforeseen at the time of the original change proposal; interferences beyond those anticipated; the owner or architect not having complied with some condition of the proposal (such as shop drawing approval time, a payment provision, etc.) or contract breach; clear contingencies that have become used up; or allowance items of the proposal that have become inadequate.

It is for these kinds of effects that the proper qualifications and reservations of rights become critical in every change order proposal. They are there to leave open doors through which any item of each finalized change can be reassessed. As soon as an effect altering the cost performance is evident, research your proposal as approved to confirm your reserved rights and any qualifications of the proposal that may have been breached.

Whether the work is proceeding on a time and material (T&M) basis or the change has not been finalized for any other reason, precise documentation is the rule that must not be compromised. All direct cost items must be kept track of, starting with the effort that went into the proposal preparation. These historical records may become important if it later becomes necessary to demonstrate just how much time the change has been consuming from other important project activities.

Whether or not the change has been finalized, it is important to realize that the documentation of the indirect and consequential costs becomes even more important to proving a construction claim. Since these are the components of every change order that are subject to the greatest scrutiny and criticism, they must be cataloged in explicit detail. Their justifications must be absolute and their calculations must be unchallengable. To accomplish these objectives, a good cost accounting system and good schedule documentation management must be in place.

Notice Requirements

Although change order proposals are not construction claims until they are refused by the owner, it is wise to make every effort to meet all stated notification requirements of any dispute resolution clause that may exist in the contract documents. Dispute clauses normally indicate a specific for notification after discovery of a change giving rise to increased time and/or costs. It is common for the owner to argue that the contractor has not met the requirement. When this happens, the owner will invariably begin the time calculation from the earliest date that the change may have become evident. A contractor's calculation, however, will begin from the date that it became known that the change will effect time and/or money. It is at that point that it will no longer be considered a job condition.

Notification can be a double-edged sword. If you strictly follow an attorney's advice on structuring the letter of notification, the language in your notification will appear as if you are ready to haul the owner into court tomorrow. This will most certainly instigate and propagate an adversarial relationship throughout the remainder of the project between owner and contractor. You'll have clarified that the working relationship is now adversarial. In all probability, you will have "protected" yourself permanently out of any informal communication, and away from an important level of trust.

On the other hand, if the contractor fears offending the owner to the point that communications become too watered down, the contractor can maintain good relations right into the red column. The balance, then, lies in the ability to formulate effective communications that may be courteous, but that also clearly protect the contractor's interests.

Cost Monitoring

Separate and distinct from the architect's design costs, the contractor must monitor project field service costs during the project. Field costs are a function of the size and classifications of the on-site field labor forces assigned to the project, field office overhead costs, materials and supplies, support services from the home office, vehicle leases and fuel charges, the field office's share of corporate general and administrative (G&A) costs, outside consultants or contract services, and job profit.

By tabulating the monthly accumulations of budget and actual field service costs and then plotting each amount on a time-versus-cost chart, similar in form to the traditional S-curve chart, you can make a visual comparison that clearly indicates not only the status of the contract at any given time, but will also show the project manager or owner any change in trend toward either a savings or a cost overrun. By also plotting a curve representing the amounts invoiced to the owner for such field services, you can provide an additional dimension of usable data to the owner.

To help prepare and maintain the chart, regular inputs are required from the subcontractors, on all field costs and hours of work in each classification at the project site. Arrangements should be made by the prime contractor to ensure that all such field data is received from the subcontractors on a regularly scheduled basis at the end of each week. Typically, the monthly pay estimates of the project's contractors' work, for progress payments, are submitted by the 25th of the month, and all submittals are in office for payment before the end of the month. Billings from the architect to the owner, however, usually are based on the closing date of the end of the month. Therefore, tabulation of these data should not interfere with the owner's review of the contractors' pay requests.

Design/Build Change Orders

A newer residential construction execution method has, in the last decade, become prominent on the scene wherein the owner contracts with a designer/builder. This means that the designer and builder are the same entity. The advantage to the design/build system, from a production standpoint, is its speed of project completion, because the best design/build execution uses phased construction as its production basis. By having components or materials inventory in advance, contractors can start the project with sections of their activities already completed.

Adding precommencement activity completions and events timing controls to the project production gives the owner the ability to jump start the project called *fast-track design/build*. However, in fast-track production it also means that the prospective bidders cannot be given a detailed set of plans and specifications on which to make their bids, as design is not yet complete. So builders bid by square footage costs added to a profit margin percentage, and contract by cost-plus contracts tied to whatever actual costs are incurred, regardless of projected costs. This quite often leads to missed or ambiguous tasks and subactivites that in turn generate change orders.

Design/build project schedules use events time manipulation to either accelerate or retard forward progress in collective or individual production phases, producing forward momentum in production scheduling. Schedule delays, obviously, throw everything off and defeat the profitability of the project. Claims are always predictable in this scenario. Time manipulation is the crux of the matter, being both the problem if used poorly or the solution if used wisely. Profitable time manipulation is applied through the proper use of free float and total float, in parallel activities as well as series activities. Adding these events' time-manipulation controls to the project schedule gives us the working foundation of fast-track. As an example of how this system will work, let's look at the basic elements of a design/build fast-track project using a simple job logic schedule of a typical home building project, scheduling the phases of that project. I have made this example bare-bones and overly simplistic for the purpose of illustration. The project's phases are:

1. Design plans and engineering calcs.
2. Approve design and plans contract.
3. Obtain building permits.
4. Bid out construction.
5. Negotiate contractors' bids.
6. Complete any required geotechnical site surveys and soils testing.
7. Negotiate project schedule with best bidder.
8. Approve and award winning general contractor's bid.
9. Obtain contractors' bonds, Workers' Comp and liability insurance.
10. Begin site work.
11. Final inspection site work.
12. First progress payment.
13. Begin foundation work.
14. Final inspection foundation work.
15. Second progress payment.
16. Begin framing work.
17. Prefinal inspection framing work.
18. Third progress payment.
19. Begin electrical, plumbing, and HVAC rough out work.
20. Final inspection framing work.
21. Fourth progress payment.
22. Prefinal inspection electrical, plumbing, and HVAC rough out work.
23. Fifth progress payment.
24. Begin interior work.
25. Prefinal inspection interior work.
26. Sixth progress payment.
27. Final inspection electrical, plumbing, and HVAC work.
28. Seventh progress payment.
29. Final inspection exterior and roof.

30. Eighth progress payment.
31. Finish out punch list.
32. Finish out changes and claims.
33. Final total building project inspection.
34. Project close out.
35. Contract close out.
36. Final payment and release of retention.
37. Obtain release of liens and file Notice of Completion.

These 37 activities could be easily shown on a bar chart, sequenced, and computed for duration of activity originally estimated and actual duration remaining. Remember, this breakout of durations completion percentage represents only the first stage of a design/build schedule. The second step is to put the phases of activities, along with the early and late start and finish events, into the formula. The 37 activities can easily be scheduled with a manual bar chart or a time-scaled logic diagram, produced on a PC with CPM software.

With a preselected contracting plan developed by the builder to start preliminary activities prior to commencement, we get the combined elements of fast-track project production. The example time-scaled project schedule now has the advantage of a head start *prior* to project commencement. The disadvantage is that the owner assumes most of the cost-performance risk of a cost-plus contract with the prime contractor. But that risk management can be handled with a preselected contracting plan of a guaranteed cost-and-profit-maximum-plus incentive contract for the builder.

Once the durations percentages reach the 34 to 45% completed design point, the contract could be let out to bid in a fast-track design/build project execution. Now our production schedule contains job logic, time manipulation tools of early and finish events, all the necessary elements of a production network diagram with activity interdependence linked by a time scale, plus the head start of phased construction activities.

While the design/build project execution system will be in wide use throughout the construction industry for the next century, new variations of phased construction project scheduling are already becoming alternative construction management options. These are the next wave of contract opportunities for the claims-conscious contractor and include "just-in-time," process-oriented modular projects, and "bridging" project-scheduling methods, which are variations of phased-construction CPM, being pursued by builders, developers, owners, architects, and engineers nationwide at an ever-increasing rate.

Phased Construction

In phased construction, components of the project are started prior to main project commencement. The big advantage to the owner in phased construction is that production of essential components of the building can begin before the total project design is completed, thereby incurring less interest cost on the construction loan by completing the project sooner. An owner who is not yet due for the first draw payment from the bank does not yet have use of bank construction loan funds, and will have to draw on personal credit lines to cover initial costs of commencing the project. This additional debt load taps many owners thin for project commencement cost contingencies, and further reduces the owner's credit line to cover those contingencies. Through phased construction, however, construction can begin with certain activities completed and ready for initial draw progress payment from the bank's construction loan funds. The project is completed sooner, the owner has income from the capital project sooner, and the construction loan is paid off sooner at a lower cost. Same project, but more income and less cost. This type of project execution equates to sizable profit to the owner, but it must be remembered that the economic viability of phased-construction project execution depends on the amount of the design completion available when the project schedule is being developed.

Phased construction has proven itself to be a viable project-execution option in multiresidential construction, and is being touted as the next wave of the future. Phased construction has also proven itself to work extremely well in other types of similar multilevel types of construction projects. The primary advantage in the phased-construction method is that construction can begin while design is still being worked out (theoretically). This should mean that the project will be completed sooner. Obviously, developers like phased construction as it makes them money and cuts their debts sooner. It will no doubt continue to be a real contender in the field of future construction project executions. Phased-construction project scheduling has evolved as an efficient building-scheduling system for obtaining equal or higher-quality construction at the lowest possible price, and in the quickest turnaround time frame. These types of nontraditional project production execution methods are indications of the changes yet to come for the professional American builder.

The economic advantage of phased construction is the cost control over the selected paths of critical activities running in series, and the many paths of activities running in parallel on the schedule. In addition, the early precommencement starts of the phased construction provide financial advantages to a project owner on a phased-construction project schedule, ideally reaching the earliest possible strategic project completion date. In phased construction, the overall responsibility for controlling project execution remains not with a builder, but with the owner.

Typically, responsibilities for completion of the project design and specification are delegated to the architect or design professional, modular components and long-lead item procurement, and construction activities to the builder, and production schedule development and tracking to the owner. Mutual cooperation between the design professional, owner, and the builder is essential throughout the project in this type of project execution. The builder and subcontractors make their field schedules in close cooperation with each other. This early builder cooperation with the architect and subcontractors produces a smooth and effective design-construction interface, and further ensures a viable and cost-effective production schedule, thereby maximizing the potential for phased-construction project profit. Phased-construction projects have been successful in major projects because they create a combined-resource approach of an integrated owner/designer/construction team, which is why phased construction is becoming a major construction project productivity enhancer.

One of the main differences in phased construction as opposed to standard execution is the handling of component construction. In phased construction the builder accelerates the delivery of critical framing and finish components, such as interior and exterior wall panels, to the project site. These are now scheduled to intersect early construction phases at key activity starting events, thus accelerating the required components and long-lead items field delivery dates along with the activity's early starting event. Early quantitative survey material takeoffs also permit construction procurement personnel to order critical bulk materials in advance, such as lumber, pipe, and fixtures. Those activities are especially important in phased-construction project execution.

In phased-construction execution, preconstruction input by the builder and owner team begins while the architect is still in the design phase, to prevent design changes later in the schedule. Builders then compensate for not having detailed finished set of plans and specifications to make their bids with cost-plus contracts tied to whatever actual costs are incurred. Activity subcontractors' bids are usually figured by square-footage takeoffs, tied to whatever finished design costs are incurred regardless of projected costs. The main phased-construction advantage in project execution is speed of project completion. This contracting environment requires early, close, and formal coordination by the owner between the builder, architect, and subcontractors. Contract conditions conflicts invariably arise in this type of project execution because of a lack of clearly assigned responsibilities.

For a phased-construction schedule to be viable and operational, the plans and specifications need to be 30 to 45% completed before the project schedule is developed and executed. Because the project schedule must interlink with other major project considerations, such as design and long-lead items procurement, getting a phased-construction's jump start without adequate finished design can incur costly construction claims. In fabrications with long-lead delivery times, error is obviously probable if the groundwork has not been done previously to provide accurate components details and site activities duration.

Design Matrix Change Orders

The most sophisticated (and currently emerging) high-end residential construction design systems incorporate what is termed *design matrix networking*, in which specialized production designs are grouped in phases and activities, then utilized in a variety of project-oriented configurations. The designer/builder works with small teams from the various areas of specialization in the project, toward common ends of interlinking completion events.

This is a highly responsive and event-oriented operations system, producing a design network that will interface through the divisional factors (matrix) in the project's design, thereby reducing time required to manage multiple production schedules. This is also one of the basic elements in CPM fast-tracking. However, design matrix networking requires sophisticated manufacturers and builders with good communication skills and a commitment to the project's quality. Missing or inadequate specifications are typically at the root of change orders in this area of contract change order prospecting. To find these potential design matrix change orders, the prospecting is a process of first determining:

- The tasks and activities to be performed
- The time, in events, to be devoted to each
- The exact specifications of materials and equipment
- Long-lead items
- The job logic sequence of each activity
- An optimistic and pessimistic completion time frame of activities and materials/equipment interlinking

Then, by combining these four tasks into a time line matrix we cut down on some of the data-processing time on most of the schedules if we are running multiple projects and multiple schedules, and if those schedules are similar. Controlling all the projects' production by design matrix networking is a combination of planning and presetting operations that include production phase milestones. The advantage to using design matrix networking is that if you do so, the summary reports (called sorts) become evaluation mechanisms *before the fact*. This gives the builder an edge in production profit, just as fast-tracking does.

This type of project production controlling is an area of opportunity for most small- and medium-sized high-end residentials running budgets of five hundred thousand to a million dollars (that is currently a small-sized high-end construction project in California, according to SBA statistics). The development of cost-efficient design matrix processes for achieving production expectations, measured by milestone indications of success, coupled with remote (field) data collection systems linked to a host processing computer back at the office, is fast becoming the difference between success and failure in this arena.

As the design matrix grows in complexity, the functions of controlling production through design matrix networking also become increasingly complicated. Accordingly, these are the areas to look for generating change orders. At this point, it is critical to keep in mind that it is not necessary for you to be a statistician to begin analyses of these time-manipulation factors. Remember, that's what the computer is for. Set up the software spreadsheet formulas to do the range statistics calculations for you. Spread your databases over many ranges and remember to leave a blank row or column between each database for additional safety in data retrieval. The big picture is what's important, and where you need to stay focused. So the more you understand about the limits of your project and the elements which define as well as constrain the project's production events, the better you will be able to manage time manipulation of those events.

Named Subcontracts

Named subcontracts, or owner-selected subcontractors, describes a case in which the special procedure for subcontract selection or the subcontractors themselves are directly handed to the prime contractor by the owner. Elaborations of this idea also include the practice in which the owner receives the individual subbids for selected categories. A list of the qualified bidders, along with their bids, is then published. The prime contractors are then to incorporate the (presumably lowest) qualified bids into their general bids. On the surface, this arrangement seems to serve the owner by providing the maximum advantages in competitive bidding. This procedure has been used with widely varying degrees of success by different public agencies, but the form has become increasing unpopular in residential construction.

Owners use this approach generally for two reasons. First, the reason (claimed) is to reduce or eliminate the practice of "chiseling" the subcontractors by prime contractors. The logic presumed that by giving the subcontractors the opportunity to submit prices that they were confident would be accepted, the subcontractors would allow their prices to become more competitive. They wouldn't have to build in "room" to negotiate.

The second reason comes down to an attempt by the owner to exert some (at least perceived) control over who will be performing what work, as well as the contractual provisions that will govern them. Typically, the arrangement will go as far as to dictate the precise terms of the agreement between the subcontractor and the prime contractor.

There is a serious difficulty (for the owner) with the whole arrangement, however. It is that all this owner activity removes the prime contractor's responsibility (and liability) for the respective subcontractor selection to a very great extent. When the subcontractor's performance degenerates, the prime contractor dutifully turns to the owner and says "What do you want to do with *your* subcontractor?" The owner will then (at least initially) revert back to the old boilerplate in an attempt to convince the prime contractor to the contrary. But after all, the owner wouldn't even let the prime contractor use its own form of subcontract. The prime contractor couldn't even determine the scope of work to be included in the respective bid package. This situation needs to be backed-up with a document similar to Figure 3-1.

The simplest kind of problem that can develop under the circumstances is when the lowest apparently qualified bidder suddenly (for whatever reason) becomes unavailable after you've used that company's bid in your own general bid. Because it was originally the owner's selection, and a simple mathematical relationship, it will generally be a straightforward matter to secure a change order that will bring the next lowest bidder on board. Straightforward, that is, for the net difference in price between the two prices only, and not so straightforward for consequential or interference costs.

Changes involving deficiencies in owner-defined work are generally more difficult to secure owner recognition. This is particularly true throughout the gray areas left at the interfaces between two or more owner-defined subcontracts (is the work described in both places, or not at all?). For example, on one project, the CSI (Construction Standards Institute) Division 16 specification described the underground electrical duct bank to be "encased in concrete in accordance with the requirements of Section 03300." Does this mean that the electrical contractor places the concrete, using Section 03300 as a guide, or does it mean that the Section 03300 concrete contractor is to place the concrete? This involved two owner-selected subcontractors. The owner ultimately paid the electrical contractor thousands of dollars to provide the concrete.

"Duplications of Design" and "Inadequate Level of Detail/Missing Details" develop this idea further. Often a disclaimer exists in the front-end documents or in the contract itself that tries to limit or remove the owner's liability for the owner's subcontractor selection. Depending on the actual language, however, it may actually help your case. For example, one specification stated that: The (specification) divisions and sections do not operate to make the owner or the architect an arbiter to establish the limits to the contracts between the general contractor and the subcontractor. If you're lucky, the language might be expanded to include: except in the case where pre-filed (owner-selected) subcontracts establish such limits.

If the complete paragraph, or something similar, is included in the General Conditions, you'll be strongly supported in your contentions. If just the first portion of the paragraph is included, your rights will be generally the same, but you'll have to be more persuasive.

LETTERHEAD

Date

To*: Subcontractor named in owner response* Re: *Project or Job number*

SUBJ: *Description used in owner determination letter*

Mr. (Ms.):

Attached is the (*date*) (*owner's name*) determination of (*subcontractor name*) responsibility for the subject work. The work is required to be complete by (*date*) to avoid (*additional*) schedule interference and associated charges. Accordingly, please immediately:

1. Confirm acceptance, and
2. Advise this office by (*insert date*) of your detailed schedule for completing this work.

Sincerely yours,

Project Prime Contractor

cc:
Owner's name
Architect's name
Subcontractor's name

File: *your computer file name/number for the project*

Figure 3-1
Decision Directing Work Form

Questionable Work

If, in this scenario, the prime contractors runs into undefined work by a specific subcontractor, the prime must submit a Letter to Owner Regarding Responsibility Determination for Questionable Work. This document is designed to notify the owner that by all appearances, the work in question is not apparently included in any of the owner's subcontract scopes. It maintains that nobody intends to do the questionable work without being directed to and reminds the owner that the contract assigns to the owner the responsibility for making the final, precise determination.

It should request a definite reference to a precise specifications section. This may become an extremely important piece of information in the event of a subsequent appeal of the owner's decision. It should specifically ask that the responsible party be determined by name and state that it hereby notifies the owner that interference will occur by a certain date without proper, timely response.

These notifications and requests for precise information by exact dates will become important basis for subsequent action by all parties. If the owner's response to the letter is less than complete, it should trigger an immediate reaction on your part. The owner must then be notified of the resulting interference. Figure 3-2 shows an example of this document.

LETTERHEAD

Date

To: *Owner or architect, per specification direction*

Re: *Project number*
Job name/description
SUBJ: *Brief description of work responsibility determination*

Mr. (Ms.):

Per our conversation of (*insert date*), both (*name subcontractor 1*) and (*name subcontractor 2*) take exception to the responsibility for (*brief description of questionable work*).

General Conditions Article (#) directs that the owner (*or architect, as described*) interpret the requirements of the contract. Accordingly, please advise of the party responsible for the subject work, indicating the applicable specification section and responsible contractor by name.

Your complete written response is required by (*insert date*) to avoid (*additional*) interference.

Sincerely yours,

Project Prime Contractor

cc:
Architect (or Owner)
Subcontractor 1
Subcontractor 2

File: *your computer file name/number for the project*

Figure 3-2
Questionable Work

To completely follow-up on this document you must determine if the agreement or the scope of work for a subcontractor has been owner defined. Then identify any disclaimers or other qualifying language that exists in the owner-prime contractor contract. Remember that despite first appearances, the language may actually help. Do the best that you can to get the owner to determine and confirm exactly what specification section the work (supposedly) is included in. Be specific. Get the owner to state the precise specification and to name the subcontractor "responsible." Then advise the named subcontractor that the owner determined the work to be that subcontractor's responsibility. Accordingly, direct the work to proceed. Send the owner copies of all correspondence. Make all directives in the owner's name.

When the subcontractor objects, advise the owner. Avoid stepping directly in the line of fire between the two. Remember that you're only there to coordinate the owner's directives with the subcontractor's actions and back again. This is an important concept that should become a way of life for the general contractor. In this way, liabilities for both the owner's and subcontractor's actions will stay where they belong. Keep them there and keep up the momentum. Immediately submit a detailed change order proposal. If it is possible to incorporate the properly substantiated proposal along with the subcontractor's objection, do it.

Chapter 4
Project Schedule Claims

Project Schedules

All construction projects are complex to some degree. For a project to be completed on time and on budget, all work (which is broken down into tasks, activities, then phases of construction) must be carefully planned and scheduled in advance. The prime contractor then should "cook" the schedule, looking for any long-lead items and all probable production bottlenecks.

A project schedule is the contractual network diagram of the project's planned activities, their sequence determined by job logic, the contractual time in working days required for completion (activity duration), and the conditions necessary for their completion (contract specifications). It is also a contract document linking the lender, owner, prime contractor, and subcontractors. It serves, within the contract specifications, to advise the lender and owner of any unsatisfactory progress in any activity's production. Any such advisory special meeting should have back-up documentation similar to Figure 4-1 as a claims management strategy.

Project scheduling always contains a duality that is a strange combination of clean, conceptual, theoretical planning which then must be carried out in a real-world application production that is sometimes akin to a fistfight in a mudhole. Seldom are the two conducive to easy production—not for anything larger than, say, a doghouse. (Unless, of course, the doghouse is located in a flood plain or earthquake area. Then, of course, engineering plans, environmental impact studies, and building permits will be required.)

Project scheduling fundamentals require that each activity within the project has its own time frame and usually cannot start until preceding activities have been finished. There may be other activities, however, that can be carried on simultaneously because the activities are entirely independent of one another,

LETTERHEAD

Date

To: *Owner or architect, per specification direction*

Re: *Project number*
Job name/description
SUBJ: *Brief description of work responsibility determination*

Mr. (Ms.):

Per our conversation this date, a special meeting will be held at the job-site on (*date*) at (*time*) to resolve the subject change. Each issue on the following agenda will be reviewed:

1. (Description)
2. (Description)
3. (Description)

Messrs. (insert the names of each person required) are required to be present to resolve these items.

Thank you for your consideration.

Sincerely yours,

Project Prime Contractor

cc:
Architect (or Owner)
Subcontractor 1
Subcontractor 2

File: *your computer file name/number for the project*

Figure 4-1
Special Meeting

such as plumbing and electrical. "Dovetailing," or interlinking, of these activities and their duration start and stop dates efficiently requires skillful time management planning in the project schedule. If such skill is not applied to producing and tracking a project schedule, change orders are bound to occur. A typical residential construction project involves many varied activities, and tasks within those activities, that are interdependent with one another as well as other activities that are totally independent of each other. When the production phases, showing all the activities, are interrelated on a project schedule bar chart, they create a multilane highway of activities moving towards completion. It also shows their individual time durations, as well as their relationships with their preceding and succeeding activities. When all the project's tasks and activities are superimposed to provide the big picture, it quickly becomes evident to the observer that project scheduling is a difficult and complicated project management function.

The project schedule must be the same one that was actually used to build the project. Even if a schedule that is substantially different from the one under consideration had been formally submitted and approved by the owner and/or the design professional, the schedule that was actually used and depended upon by the various trades will generally be considered to be the contract document. Any special meeting must be documented. (See Figure 4-1.)

The project schedule must be periodically revised. It is generally recognized that project scheduling, however exact or inexact, is nothing more than a plan to construct what can reasonably be construed from the contract documents, considering that there may be several ways to build the same project. Although contingencies may be included to allow for imprecisely defined variables, there probably never has been a schedule that anticipated every problem, coordinated every piece of work, and required no modifications to make it work. Changes to, or corrections of, the plan are inevitable. The schedule, therefore, must be updated periodically to maintain a current and accurate representation of reality.

The periodic updates of the project schedule must show all positive and negative influences by all parties. None must be singled out; none must be absent for convenience. If all updates, for example, only indicate delays caused by the owner and fall to delineate other known problems caused by other parties, it won't be very difficult for you to demonstrate that bias has been built into the document. Failing to recognize all significant events will bring into question the validity of the entire project schedule presentation.

The project schedule must include realistic construction logic and activity durations to demonstrate the professional ability of the planning and scheduling team. Illogical sequences or the lack of consideration of critical variables (such as indicating the installation of HVAC units in a area with inadequate access) will only demonstrate that the constraints in the project schedule were unrealistic to begin with. The project schedule must fairly represent the actual method planned to build the project. If, for example, the project schedule had been prepared primarily to cater to progress payments, or was otherwise unrelated to actual construction of the project, it would become clear that, as a tool for managing the project, its value was marginal at best.

The project schedule updates and analyses must be realistic and in perspective. An overly aggressive computation of damages might hurt the validity of the entire analysis. Direct cause-effect relationships must be shown. In addition to the foregoing, the project contractor must be sure that all notes and references are correlated with the more detailed, chronological project correspondence. Have all field notes, claims, commitments, deadlines, and promises detailed in the schedule documents. Include the dates and the names of the individuals in those organizations supplying information. Pin it down.

Job Logic

The logical sequence of the project's construction activities, factored by local practical limitations, is referred to as *job logic*. The project schedule's activities chosen may represent relatively large segments of the project or may be limited to only small steps. To use an example, a concrete slab may be a single activity on a small job, but on a larger job it will be broken into the separate steps necessary to construct it, such as excavation, sub-ex preparation, erection of forms, placing of steel, placing of concrete, finishing, curing, and stripping of forms. As the separate activities are identified and defined, the sequence relationships between them must be determined (see Figure 4-2.) These relationships are referred to as job logic and consist of the necessary time durations and sequential order of typical local construction operations that are unique to your geographic area.

It is a fundamental in project scheduling that each activity must have a determined starting event, which may be either its own start or the finish of the preceding activity. Activity durations cannot overlap their finish events.

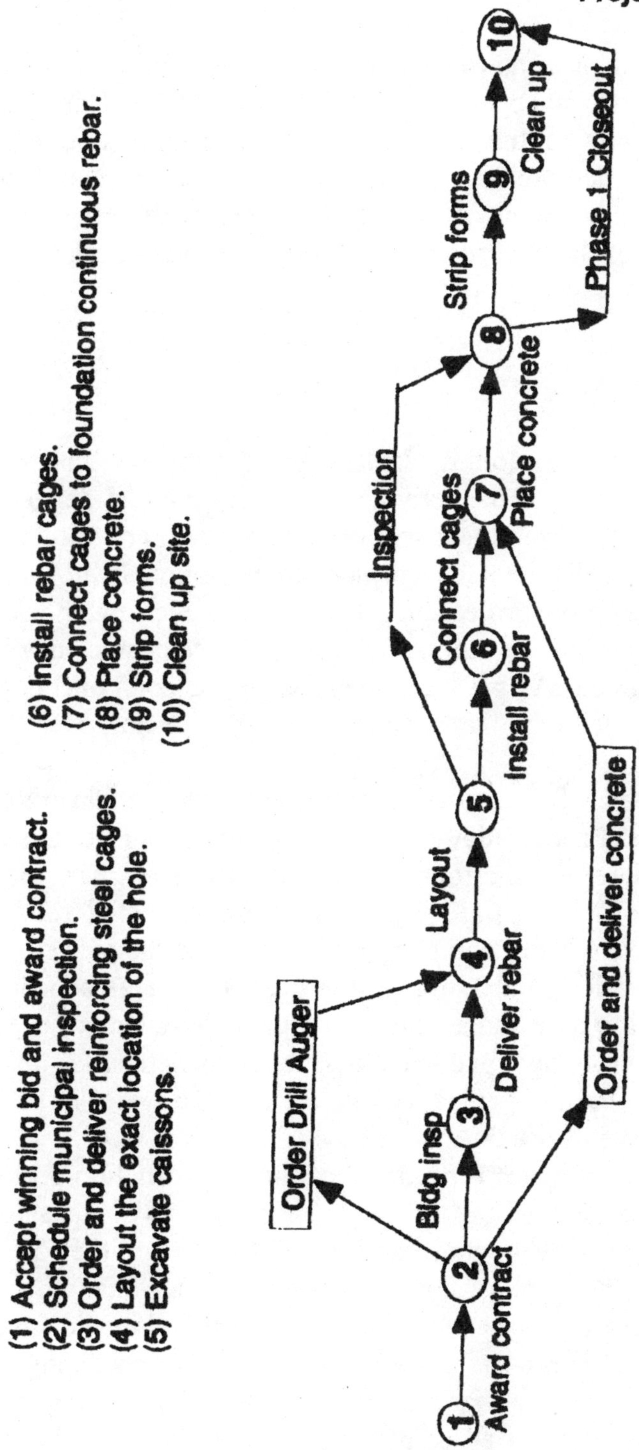

Figure 4-2
Job Logic

Therefore, job logic is established to provide operations sequences within practical constraints. Established job logic is then used to build program logic (see *Program Logic* later in this chapter) within the computerized project schedule program. By determining the job logic, activities can have their interdependencies critically examined during all phases of the schedule, before errors occur and result in change orders, causing delays and costing money.

Scheduling Fundamentals

A construction project only becomes viable when the project schedule has a functioning network diagram of the activities, components, and delivery dates. A working flowchart with all job logic sequence developed, and total durations required for completion added, creates a project network time line. A schedule's primary purpose is to keep the project on schedule. However, much useful data from the schedule can be used in other ways for added cost-efficiency in many areas of project production. As we have seen, panelized components increase productivity, efficiency, and time management, thereby reducing overall project cost.

The fundamentals of construction scheduling can be illustrated by a simple project, such as a room addition. The activities could be as simple as starting and finishing dates for the foundation, walls and roof. The foundation and walls, once site-constructed, are topped with prefabricated roof trusses, finishing the job quicker. A more complex project, such as the entire residence, will add phases of activities completion. These phases serve as benchmarks in construction completed to date and according progress payments to the activity subcontractors. In scheduling fundamentals, these benchmarks of phases completion are called *milestones*. Financial milestones are typically tied to the dates that progress payments are due.

Bar charts are the simplest form of scheduling construction projects and have been in use the longest of any of the systems we have available today. They offer the advantage of being cheap and simple to prepare, they are easy to read and update, and they are readily understood by anyone with a basic knowledge of the construction project business. They are still in wide use today, even as one of the final sort reports of computerized CPM scheduling systems. The main disadvantage of the bar chart is its inability to show enough interactive relationships between all of the site-built and modular components activities on larger, more complex projects.

Simple job logic of placing the activities in order of their construction sequence is used. Normally, many activities have been condensed into single group bars to get the complete chart to fit on one page. However, any time you condense something by compressing it into something else, you lose control over that item.

In project scheduling, the approach is just the opposite. We "explode" the activities, subactivities, tasks, and work items out into a network schedule so that each has complete computer tracking and controlling. Each of the macroactivity bars in standard bar charts need to be broken down into a series of individual tasks, to control the microactivities within the macroactivity. Only by having control at those levels will we have ultimate control of each macroactivity.

Accordingly, a network bar chart will show macroactivity progress, which is useful and will suffice on small jobs, but no interrelationship between the modular and site-built activities that would allow for alternate routes if site-built progress hits a bottleneck. This ability to switch paths is crucial to critical path management (CPM). It's much like tap dancing in a minefield: You can't stop moving or you lose. If project production stops, everyone loses, so it's in everyone's best interest that things keep moving to a timely completion. Bar charts have no capacity for contingency planning on the project schedule. And as a builder interested in profit, you need to be aware of the potential for this system to generate change orders by having components and construction options ready.

Scheduling control comes at the microgrouping, or individual task, level. Manually generated bar charts and spreadsheets that cover the whole project involve a lot of detailed record keeping and summary reports, which soon absorb too much of the builder's time to allow managing the field end of the project's production work properly.

The primary reason for the cost overruns and late completions of projects in the 1960s was this lack of control over the project's schedule time line, and became a focal point of change for more efficient construction claims management. Delay overruns on commercial and industrial projects can run into millions of dollars per day. Traditional site-built techniques used up to that time proved inadequate. Scheduling larger projects in that sort of project execution with standard bar charts simply does not offer the production time-manipulation advantages of modern computerized project scheduling.

Project Scheduling Terminology

To understand conventional project scheduling, you first must know the lingo of scheduling, along with the terms' meanings in network diagram scheduling. Here are the principal terms and definitions of procedures of network scheduling, in order of their precedence:

Task. A task is an individual unit of work that may combine with other tasks to complete an activity or be independent work items. Tasks can be thought of as the separate work unit items that collectively need to be done to finish each activity sufficiently enough to start the next activity.

Activity. Activities are any single identifiable work step in the total project's production. Groups of tasks, combined to finish a job item, are activities. Once the activities have been identified, their sequential task logic is established. Then begins outlining the project graphically into a network diagram. In a velocity diagram, the symbol for an activity is an arrow. The arrow connotes linear time line movement from left to right, start to finish. The continuity indicator is the arrow from one activity to another.

Activity number. Number assigned to each activity. These numbers should be sequential. The computer will not be able to establish job logic if activity numbers are nonsequential.

Activity list. List of work items for a project; also the work break-down structure.

Activity duration. Elapsed time to perform an activity, start to finish.

Arrow diagram. Network diagramming method using arrows to show activities interrelationship and the flow of job logic. In network diagramming, the symbol for an activity is an arrow. The arrow represents linear time-line movement from left to right, start to finish. The production continuity is the arrow moving from a preceding activity to a succeeding activity. Arrows in a velocity network diagram represent the activity itself, not the direction of movement. The angle of slope or the arrow's length are not factors in the scheduling, simply the designer's choice. Each arrow in the velocity network diagram represents either an independent activity or an interdependent activity. Each arrow is identified by its respective activity numbers.

Dummy arrows. Dashed-line arrows in scheduling diagramming are the symbols for dummy arrows, which are diagramming symbolism showing the constraint dependency between activities. Any dummy activity acting as a constraint is shown as a dashed line with zero elapsed time. The direction of the arrow shows the production order of activities, and a dummy arrow shows what activities constrain the start of another activity or activities. To overcome lack of separation and control, the builder can use dummy arrows to assign *i-j* numbers to each activity, thereby giving each activity its own computer cell address. Dummy arrows are also used on the project schedule to assign different *i-j* numbers to any activities that are running in parallel.

Event. An event is the exact day at which an activity is just starting or finishing. Network program logic applying to all events is that all activities leading into an event can be started at that time. An activity is always preceded by an event and followed by a sequential event. Thus, an activity always has both a starting event and finishing event. Theoretically, that finishing event is the starting event of the next activity. However, one must always account for the natural constraints of the real world and provide a timing-leeway cushion allowance, as problems in actual production will require the ability of contingency planning for tasks and/or activities that may get bottlenecked along the way. Proper use of float is the tool in the method of contingency planning here, which is known by the term *workaround*.

Event diagram. The most common event system uses circles at the ends of each activity arrow. These circles are placed at the junction of the arrows and they represent the event, or the moment of time at which an activity is just starting or finishing. Important events which mark the completion of a phase, such as a foundation final, are called milestone events for they serve as benchmarks to your schedule's progress versus run time.

Milestones. Date on the schedule predetermined for a phase or important occurrence is scheduled to take place. Milestones are benchmarks of long-term and short-term progress events that require some reasonable progress in activities or tasks toward a desirable outcome. Milestones are specific, time-limited objectives which, in combination, are sufficient to achieve those progress goals. Milestones are placed in the schedule to produce:

- Motivation for project team, builder, and subcontractors
- Utilization of competitive advantages
- Repair schedule competitive disadvantages
- Acceptable progress to the owner and project team if those milestones are met on time and within budget for progress payments to be paid
- Measurement of actual progress versus scheduled progress

At the stage of the process in installing milestones, the builder should be thinking in broad terms. As one studies particular areas of the schedule, one develops progress objectives that can be related as milestones. Milestones should reflect both strategic planning as well as current period objectives. They should also represent success, not the absence of failure or the remediation of existing problems. This is an important concept for the professional project builder to grasp and use in developing the schedule. Unscrewing a problem isn't progress if the problem should not have happened, so milestones should mark progress only. Positive benchmarking is important; milestones should be specific and measurable objectives. They should also appear as projections of progress evaluations at the end of each construction phase. Interim milestones should measure phases of progress which are critical to meeting long-term project timing. They should "back down" from the specified completion date to the present in sequential order.

Once you have identified milestone phase areas, schedule dates for those events and the completion objectives that go with them. If you are scheduling with a group, such as the project team, it is important to agree upon the goals and objectives of each phase, leading to the establishment of milestones as benchmarks to evaluate the progress towards those goals and objectives. It is necessary to agree upon milestones prior to establishing phases because milestones should lead the completion of each phase.

To reach milestone events on time, it is necessary to back down to the present. If you are to reach the critical completion date, where must the project be in one month? In two months? Next quarter? To actually achieve your milestones, it will be necessary to periodically re-evaluate start and finish events and use float to close gaps in timing. In planning the schedule, don't spend unreasonable amounts of time on making "perfect phases," but instead create obtainable and realistic milestones by which you can measure the project's progress.

The best planning process is to estimate durations based on what you know and your company's historical data. Include in the schedule periodic sorts, or summary reports, that will test the milestones objectives for the coming quarter. This tactic will improve future milestone projections. Remember, the purpose of planning is to reduce risk. The best schedule plan is one that allows you to test your estimates before they cost lots of money. This is always the owner's perspective and should be yours as well. We builders do this by running "what-if?" scenarios, varying activity early and late starts, and varying the activity early and late finishes. A test of the schedule's quality is the extent to which completion events are balanced against float. Every project schedule needs to be built upon this basic criteria:

- Each macroactivity step, defined into phases. This should be detailed enough to be measurable and give guidance to those who are responsible for managing that area of work.
- The person responsible for seeing that the activity is completed. This is usually the prime contractor, the site supervisor, a project team member, or an activity subcontractor.
- Activity cost in relationship to total budget. This should include time estimates, indirect costs, and direct costs. In the budget sort, a calculation needs to appear with this data to build a cost-tracking analysis.
- Project phases deadlines. These should include the actual time available by the subcontractor responsible for the activity, including its assigned early and late start and finish events.

All of these four steps should be closely examined for potential change orders and claims, up to and including project closeout. In larger and more sophisticated projects, this basic outline of plan elements will be more complex but will still follow these basics in a larger configuration. Time and cost estimates are made in greater detail for each activity. "Event dependencies" are established, which must be accomplished before others can be begun. Resources that must be devoted to each are listed, then broken out from the total budget. Resources are then allocated with priority weighing, to ensure that critical path activities are resource-covered first. Progress-checking milestones and specific sorts are built into the procedure to ensure project schedule compliance.

Early start date. Earliest date an activity can start.

Late start date. Latest date an activity can start.

Early finish date. Earliest date an activity will be completed without float.

Late finish date. Latest date an activity will be completed without negative float.

***I-J* number.** The letter *i* designated symbol for the tail of an arrow (start of an activity), and the letter *j* designated symbol of the head of an arrow (finish of the activity). In CPM, events are the exact day an activity starts or finishes. They are also dates of milestone completions. Events are assigned identification numbers for computer processing. The starting event number is the *i* number, and the completion event is the *j* number. The *i-j* number is used as a relative cell address for recording the activity's data. If CPM diagrams were to be prepared using random activity numbering, all activity numbers in the entire network would have to be renumbered to allow any new or changed activity to fit in a sequence with the other activities. Therefore, it is a basic CPM requirement that, when event numbers are assigned, the finishing event number at the head of the arrow must be greater than the starting event number at the tail of the arrow, and the *j* value of each activity must be greater than its *i* value. A typical CPM network can involve hundreds of separate activities that must remain flexible in scheduling, so the experienced project builder assigns the activities *i-j* numbers only *after* the entire network has been completed and is ready for its first "cooking" or trial-run computation.

Using the vertical and horizontal axes of graph coordinates, *i-j* events can be displayed in either plane. The vertical numbering method is more widely used. It numbers all events in a vertical column in sequence from top to bottom that equates to a parallel time line. Those groups of activities then move from left to right. There is no real significance to the event numbers themselves except as means of identifying activities, so if the CPM format of keeping the *j* value of each activity greater than its *i* value is used, blank cells can be left in the numbering system so that spare numbers are available for changes or additional work that may come up. Sequential *i-j* numbering provides this flexibility in scheduling while also providing the computer with program logic data for events and activities locations on the network diagram.

Float. Measure of available scheduling leeway time on any activity's completion. Float can best be described as scheduling leeway or slack time between production activities. In CPM, float is the number of days a noncritical path activity can be delayed before it becomes part of the critical path. All float types that we will look at in this chapter are in increments of working days. It is in the best interest of the prime contractor, subcontractors, the project management team, and the owner that the project have float periods. These float periods assure that every activity has some time flexibility. This window of cushioning can be opened or closed (accelerated or constrained) at a critical point to provide a cost-effective production time frame. Float, properly used in CPM scheduling, is a major tool of time manipulation.

The float of an activity represents the scheduling timing leeway available to be committed for total project critical path duration. When total float is available, the earliest start of an activity can be delayed, its duration extended, or a combination of both can occur as long as the late finish time is not exceeded. It is the window of timing that is available for that activity if the activities preceding it are started as early as possible and the ones following it as late as possible. When an activity has free float time available, this extra time may be utilized to accelerate or constrain succeeding activities. Critical path management uses float as a time manipulation tool, and the nomenclature and workings of float comprise essential knowledge of the builder.

Float periods show that every project schedule, even fast-track, has some time line flexibility. A noncritical path activity can be started a few days after it was scheduled to begin without delaying the critical completion date, whereas a critical path activity must be started immediately once the preceding critical path activity has been completed. Additionally, a noncritical path activity can finish later than the date it was scheduled for completion. However, if a critical path activity's completion is delayed, the total project will be delayed. In terms of constraint on a noncritical path activity, it cannot be delayed longer than its float period without moving to and becoming part of the critical path. By using float in CPM scheduling, time manipulation of noncritical path activities means accelerating or constraining their starting events a few days after they were originally scheduled, without delaying the total project's completion date.

When an activity has float time available, these extra days may be used to serve a variety of scheduling purposes. When float is available, the earliest starting event of an activity can be delayed, its finishing event extended, or a combination of both can occur. We do this through constraints and acceleration. Constraints are explained later in this chapter; acceleration speaks for itself. If we can string activities tightly together, obviously we'll save the most money. To successfully manage the schedules of noncritical activities and work items within the critical activities on the critical path, the project builder must understand the interdependent workings of float times and constraints.

Free float. Time by which the activity finish event can be delayed without affecting the succeeding activity's start event. There are three basic types of float time and other subfactoring types of float within those basic three. The major three are: free float, total float, and negative float. Free float is the exact amount of days that any specified activity can be delayed without delaying the early start of the succeeding activity. As we learned earlier, events mark these days and therein the days of free float available on that activity. Free float is the least difference in days between the early finish event of an activity and the early start event of all following activities. The total float of an activity is the maximum time that its actual completion can go beyond its earliest finish time and not delay the entire project. If all the total float is used on one activity, a new critical path is created. The free float of an activity is the maximum time allotted by which its completion date can exceed its latest finish event without affecting either subsequent activities or the project completion. If a work item within an activity is delayed enough to consume an activity's free float, if it still completes on time, the activities following it are not affected and they can still start at their earliest start times.

Total float. Measure of available spare time or scheduling leeway available on all activities' completion. Total sum of all free float. Total float is the amount of time that an activity can be delayed without delaying the late finish event for the project's completion. Total float is shared with all activities. When an activity has a certain amount of total float, it can be used without tighter scheduling constraints affecting all the other critical path activities.

Free float, however, is not shared with other activities as is total float, and accordingly free float provides the only true measure of how many days an activity can be delayed or extended without delaying any of the other activities. It must also be added as part of the total float of preceding activities. In a network schedule, an arrow diagram is used to show a CPM network with *i-j* numbers shown in the circles and the numbers beneath or across from the arrows representing time duration required in days. Free float would be a portion of each activity's noncritical duration. Total float would be the sum of each activity's remaining free float that could be committed to the critical path. Total float is shown on the CPM summary report "Sort By Total Float/Late Start." It is computed twice by the computer; once plotted on the spreadsheet for early start and finish events, then again plotted for late start and finish events.

Line Float. Line float is the amount of time slack available per line item on a computer spreadsheet. The line float time is easy to compute because it is simply the difference between the early and late dates for an activity. It represents the available time between the earliest time in which an activity can be accomplished (based upon the status of the project to date) and the latest time by which it must finish for the project to complete by its event deadline.

Negative float. Time a critical activity is late in meeting its finish event. When the critical path has been delayed and production is now behind schedule, the earliest starting event when an activity can begin is now past the latest time in which it can be completed to stay on schedule. This is known as negative float. As the activity has no float, its completion time is reduced to critical duration, and if it is behind, the difference between the early and late dates is less than zero. Negative float shows how far behind schedule the activity is, and if it is a critical path activity, it shows how late the project completion will be.

Distributed Float. Float is sometimes used as a legal weapon by owners and builders to approve or deny claims for liquidated damages in CPM contracts. Distributed float calculations are used in network scheduling to prevent float being used up by either party in a premature or unjustifiable manner, thereby deleting float from the total project. Distributed float allocates float to each individual activity by using PERT system duration averaging.

Low Float. The most critical work activities within the critical path(s) are assigned low float timing. Leeway is crucial to contractual completion date. Low float activities are listed first on the total float sort, for primary attention.

High Float. Less critical work activities within the critical path(s) can be scheduled with more event timing leeway. High float appearing in later project activities is listed on the total float sort for schedule recycling analysis.

Constraints. A constraint is a potential real-world limitation, or a tactic of the builder in creating a float window during weekly recycling of the schedule, that can delay the starts of activities or tasks. Constraints, known as dummy activities on network schedules, are shown as dashed lines with zero elapsed time. Constraints are negative factors that balance the calculations of job logic. Constraints are delays placed upon an activity's starting event by practical limitations, preceding activity delay, or the project builder's time manipulation of that event. They are used as factoring tools in time scale computations of the network schedule. The only fundamental in constraint on a noncritical path activity is that it cannot be delayed longer than its free float available. If a noncritical path activity is constrained beyond its float period, it therefore reaches negative float and automatically becomes part of the critical path. As the project proceeds, some of the activities that were previously on the critical path will no longer be there. Suppose there is a three-day delay to site foundation. Originally, foundation had one day of float. Now the schedule must be readjusted because the total project has now been delayed two days. Foundation has now become a critical path activity, and site grading has moved off the path.

Remember, constraints are those practical limitations that can influence the start of certain activities. The activity that involves the placing of reinforcing steel obviously cannot start until the steel is on the site. Therefore, the start of the activity of placing reinforcing steel is constrained by the time required to prepare and approve the necessary shop drawings, fabricate the steel, and deliver it to the jobsite. In project scheduling, you treat constraints the same as activities and represent them as event durations on the network diagram. These types of constraints prevent activities from starting until other preceding activities are sufficiently complete. Policy constraints are those methods the owner or project team uses in the project's production to consume total float.

Constraints are shown as dummy arrows on the project schedule with dashed lines and zero elapsed time.

Phases. Phases are groups of activities that will happen in a logical order, precede or succeed one another, or happen simultaneously. As tasks are the micro units, phases are the project's macro units. Phases are arranged in divisions with sequential velocity management. Phases are supposed to flow into each other smoothly, on time, with no problems. If problems are anticipated, contingency planning in the form of workarounds need to be established ahead of time.

Job logic. This is sequential relationships between activities, identified and defined during preschedule planning. These relationships consist of the necessary time and sequential order of construction operations throughout the project. The logical sequence of the project's construction activities, factored by local practical limitations, is referred to as job logic. The activities chosen may represent relatively large segments of the project or may be limited to only small steps. To use a previous example, a concrete slab may be a single activity on a small job, but on a larger job it will be broken into the separate steps necessary to construct it, such as excavation, sub-ex preparation, erection of forms, placing of steel, placing of concrete, finishing, curing, and stripping of forms. As the separate activities are identified and defined, the sequence relationships between them must be determined. These relationships are referred to as job logic and consist of the necessary time durations and sequential order of typical local construction operations that are unique to your geographic area.

It is a fundamental in project scheduling that each activity must have a determined starting event, which may be either its own start or the finish of the preceding activity. Activity durations cannot overlap their finish events. Therefore, job logic is established to provide operations sequences within practical constraints. Established job logic is then used to build program logic within a computerized scheduling program. By determining the job logic, activities can have their interdependencies critically examined during all phases of the schedule, before errors occur, costing delays and money.

Logic diagram. Arrow diagram of complete project network schedule, or a cross section of an area of production. This diagram is structured showing the activities in sequential job logic.

Logic loops. The logic loop is a paradox in network planning. It indicates that a critical activity must be followed by another critical activity which has already been completed. Even as I reread that last sentence, I realize it doesn't make any sense, and I'm the one who wrote it. But bear with me and

I'll try to explain. The term logic loop is really an oxymoron, since a logic loop is anything but logical. Logic loops should be called illogic loops, but again, we professionals like to keep our industry jargon as confusing as possible. If, in our network schedule, one arrow was inadvertently connected to the wrong node, its path might well be shown to the computer to run backward. Although this seems obvious in a simple, single-path CPM schedule, if we multiplied the paths by hundreds of activities such as in large commercial or industrial network diagrams, logic loops can be easily overlooked in the planning stage.

In the vertical method of scheduling event nodes, which is more widely used, numbers of all events are in a vertical column in sequence from top to bottom, which equates to a parallel time line. Because of the vertical configuration, activity job logic can have logic loops in those vertical groups of activities without the builder realizing that they are there, the error then moving from left to right on the time line with the activity group.

I advise you to study the network diagram carefully at the beginning of the development of the schedule to confirm the job logic of the structure and to search carefully for logic loops. The best method to safeguard against them is to use sequential *i-j* numbering. The computer sort printout cannot indicate any clues to the presence of logic loops under the use of a random *i-j* numbering system, so random numbering guarantees a greater likelihood of error by allowing logical loops to remain undiscovered. Accordingly, to make ultimate use of computer program logic (the database, which in this case is the *i-j* numbers) we must number the activities sequentially.

Time-Scaled chart. Logic diagram with an added time scale.

Velocity diagram. Velocity diagram means a straight-line, time-scheduled flowchart. The purpose of the velocity diagram is to determine the most efficient paths to joining the activities in total operational network scope. Traditionally, the arrows in a velocity network diagram represent the activity itself, not the direction of movement. The angle of slope and arrow's length are not factors in the scheduling, simply the designer's choice. Each arrow in a velocity network diagram represents an activity, identified by its activity number.

Velocity network schedule. This is the culmination stage of CPM. All the preceding scheduling elements are computed into a master network plan for the entire project's scheduling, including post-project closeout. Here, each activity is now also assigned its relative *i-j* number.

PERT. Program evaluation research technique. Mainly used by the military and defense industry contractors. Optimistic time = Earliest completion, shortest time. Pessimistic time = Latest completion, longest time. Realistic time = Normal completion, average time.

CPM schedule. A schedule using the critical path method of activities management.

PDM schedule. A schedule using the precedence (or node) diagramming method.

Bar Charts. The bar chart was invented in the late 1950s by Henry Gantt. He was working as an engineer on a capital works job and developed the bar chart as a graphic representation of the project schedule he proposed. It was so successful that soon everybody was using them for tracking vertical components through a horizontal time line, which was applicable to every business in existence. Bar charts are now used in all industries for easy and quick visual representation of data. As construction scheduling became more sophisticated, the bar chart was used in combination with hand spreadsheets to schedule and track construction projects.

Computerized CPM scheduling also produces a bar chart printout sort, which is sorted from the activities' early starts. Bar charts are included in the weekly meetings, as progress reports of scheduled progress versus actual progress. These bar chart sorts are an essential tool for presentation of network scheduling details in the CPM project schedule to the owner and the project team.

Bar charts are effective at activity scheduling and tracking as a job logic diagram. They are cost-effective scheduling tools for smaller projects and on residential work. In commercial and industrial work, builders and site supervisors normally have bar charts covering walls of the job trailer with lots of different colors of highlighter going everywhere on the diagram. No one but an owner or builder has the time or the desire to compute the variables in the schedule.

Bar charts, however, do not show all the interdependencies between activities, whereas CPM schedules show the dependence (constraint) of one starting activity and the finish of the activity preceding it. Further, bar charts do not allow for variable float control at those activities' events. There are also limits to the number of activities, usually around 50, that can be tracked on a bar chart before the chart becomes overloaded and the milestones within the bar chart schedule miss their marks.

S-Charts. S-curve charts, known in the trades as S-charts, add time line momentum to bar charts and can be thought of as flow charts. These are the types of graphs that use coordinates to plot flowlines. These are scaled in a tighter time line than the bar chart to provide more scheduling control over subtasks within activities. Bar charts are typically structured with a one-month time line, whereas S-charts are structured with a one-week time line, which opens up the activity and its subtasks to greater detail and control. The project builder now obtains the margins between work items necessary to establish workable free float.

S-charts show the interrelationships of sequential and parallel activities more clearly than do bar charts. The manipulation of the starting and finishing events is critical to effective CPM. Even work items within subtasks within activities should be broken out, with individual computer numbers for dedicated cell address data storage. All activities are made up of smaller activities that need attention and control. For example, if we have an activity "Obtain building permits," it is entered and computed as one activity, whereas anyone experienced in pulling permits will tell you that obtaining building permits is far from a one-step operation. Many subactivities are involved in the permit process alone. Each of the activities has subactivities involved, which are not shown on the S-chart for clarity of concept.

Once the builder has identified these work items at the microcosm level, the activities listed on the primary report file (sort by activities) are then broken out from the sort-by-activities time line and transposed onto the accelerated S-chart time line. This, in effect, shifts the production network into double-time. Professional ball players have reported that, when they are really in the groove and doing well, things seem to slow down a notch and they can really control their timing. I once read that Matt Williams remarked that he started his home run streak when "time seemed to go into slow motion and I could see every stitch lace on the baseball quite clearly as it was coming at me." So he focused better and connected. This is exactly what happens at this transposition point of superimposing the network time line S-chart on the sort-by-activities bar chart. Here is where the builder must focus and connect. This is the point in your examination of the owner's schedule and the development of your schedule where the double-timing opens up details on subtasks and work items within the activities, and thus, windows of float. Superimposed events and milestones begin to interweave, and the job logic of the network project schedule becomes workable.

Once the builder has identified the job logic, the project schedule is plotted in a flow chart by using arrow diagramming. In arrow diagramming, the basic unit is a work activity that occurs between two events. The events in arrow diagramming are known as nodes. The events, or nodes, are numbered sequentially, and the activity is identified by the beginning and ending event numbers. Those numbers are also designated as the *i-j* number, as shown on the computer summary report Sort by *I-J* Numbers.

Each activity has a projected duration necessary to complete the activity. Each activity's estimated duration is determined by the parameters of the work items involved. The best source of data for factoring here is your company's historical data, available from previous similar projects. If there are no historical data on the activity, the typical builder uses cost estimate publications, such as Lee Saylor or R. S. Means Cost Index, and uses area adjustments for geographical cost factoring. Experienced builders will also seek a best estimate of the activity's duration from input from the subcontractors most experienced in performing that activity. The function of the arrow diagram is to determine the longest time path through the diagram, which is the critical path.

Performance Targets

Performance expectations are at the root of maintaining a successful project schedule. In the first instance, this is the process of setting production objectives: the percentage of activities completion per day and the relative cost-effectiveness projected versus actual cost. In each phase of production, the builder and activities subcontractors need to know exactly what is contractually expected of them. In the second instance, performance targets are set to establish milestones for phases completion. Modern construction management theory is evolving toward the position that contractors who will be affected by critical path milestone events should participate in creating those milestones during the schedule planning period. Accordingly, as part of the planning process, current standards and methodology should be reviewed periodically before the schedule is implemented. Performance targets should be concentrated around areas of progress development or production constraint, which should then be critically examined and options or workarounds considered.

The planning of any network schedule must start from a work breakdown structure (which is the detailed activity list) showing each activity to be scheduled. Sometimes it looks like an activity tree, as one major item is subdivided into its major parts. For example, we might have an activity called "install foundation," which could be further subdivided into "excavation," "forming," "setting rebar," and "pouring concrete." The foundation activity could be further subdivided into "buildings" (affecting structural steel erection) and "equipment" (affecting equipment delivery and erection). The decisions affecting the degree of work activity breakdown will determine the length of the activity list and the complexity of the overall construction schedule. As everything in project scheduling is series-strung, one thing leads to another. All decisions affect the whole. If there is ambiguity, change orders can be prospected. Conversely, the greater the detail, the greater the control over changes.

Accordingly, the first step in project scheduling is the planning stage. The project must first be broken down into its activities and modular construction options. The extent to which the project is subdivided into tasks, activities, and the project's capability for phased construction components depends upon the size of the project. However, the steps in project scheduling are always the same:

- **Create milestones**. Planning establishes expectations and goals for your project schedule. It requires projectwide coordination and project team commitment to common goals of achievements in critical activities.
- **Establish priorities**. Planning requires you to consider basic goals, policies, plans, and resources, and their relative importance. By establishing priorities, you also establish a relative scale of importance to assess your resources allotments.
- **Predict problems**. Planning predicts problems and production bottlenecks in activities and operations. It invites the creation of controls and evaluation procedures.
- **Solve problems.** Contingency planning allows you to evaluate what doesn't work (identify the potential change order), and then to consider what might be done to unscrew the situation. With a plan detailing your assumptions and expectations, it is much easier to identify what went wrong when problems arise in the project schedule. Even guessing will allow you to check and improve areas of concern within the schedule.
- **Go for profitability**. Planning forces consideration of the financial issues and owner's cost before resources are committed. It imposes a test on intuitive choices, points out options, and allows for you to consider alternatives.
- **Establish motivation**. Planning also encourages full participation from the project team and motivates everyone concerned, including you, to put forth their best efforts.

- **Create cost control**. Planning controls costs by creating a demand for periodic reporting, audit tracking, similar comparisons, and cost-effectiveness evaluation. This is a big area of concern to all owners and is one of the major advantages of having a professional builder on the project team, as architects and design professionals traditionally have shown a casual attitude towards an owner' cost control needs.
- **Encourage development**. Planning is the only way to put milestones into the operational phases of the schedule accurately. It requires considered decisions, rather than default outcomes, balancing needs and demands for resources.

If we boiled all this down into short little title tags to aid in memorizing, the list of item name tags would look like this:

Establish objectives

Organize

Develop project team

Motivate

Hire contractors

Communicate

Measure

Record

Analyze

If these steps are considered activities, some of the relative tasks within each activity would be steps like these:

- Designate different areas of activity completion responsibility, such as subcontracted work, that are separate from that work being done directly by the prime builder
- Spreadsheet list of different categories of work as distinguished by trade
- Spreadsheet list of categories of work as distinguished by equipment
- Spreadsheet list of different categories of work as distinguished by materials
- Breakout distinct and identifiable subdivisions of structural work
- Locate all work within the project that necessitates different timing
- Identify all long-lead items that will be special ordered
- Use owner's breakdown for bidding or payment purposes
- Use contractor's breakdown for estimating or cost accounting purposes
- Use weekly summary reports to track progress

A comparison of the amount of detail covered in a computerized project schedule work breakdown structure, and a comparable bar chart, clearly shows the difference in the amount of detail covered by the two systems. Bar charts cannot even approach computerized schedules in activity numbers without becoming completely unworkable. An outstanding advantage of computerized project schedules is the intangible benefit of forcing the project team during the planning stage to dissect the project into all of its working parts. This forces the early critical analysis of each work activity. The project team will then run the first pass of the schedule several times, to test and debug the logic diagram before the final version is ready for owner review and approval. A contractor's initial CPM logic diagram equates to a practice run on computer for the full-blown project schedule.

Program Logic

Program logic in the project schedule can best be described as the way the computer puts together the time sequence for the activities involved in the project. The program logic for the sequence of operations in this example of "Installing Caissons," would be based on the following job logic:

Activity	Sequence	Symbol
Accept winning bid and award contract	1	BC
Schedule municipal inspection	2	mi
Order and deliver reinforcing steel cages	2	OD
Excavate caisson	3	EC
Install rebar cages	4	IR
Connect cages to foundation rebar	4	CR
Place concrete	5	PC
Strip forms	6	SF
Clean up site	7	cu

This is the type of data strategy that is decided by the architect initially, then by approval of the project owner in the planning stage. For the contractor, job logic requires that each of the activities in the network has a definite event to mark its starting point and another to mark its completion point. This event may be either the start or the completion of preceding activities. See Figure 4-3.

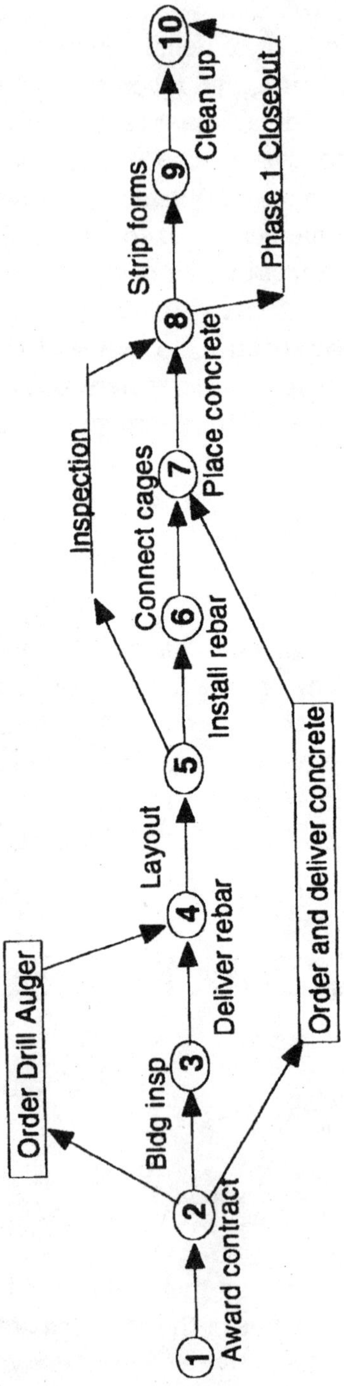

Figure 4-3
Program Logic

Preconstruction Planning

At the initial stage of the preconstruction planning, your agenda should be to compile a complete list showing all milestone events and their respective dates. These should include all meetings, prime and subcontractors progress payment schedule, progress inspections and testing requirements, equipment delivery dates, long-lead items delivery dates, submittals, and project phases completion dates. Job logic is examined here for signs of possible change orders.

As contracts are awarded, all the subcontractors should be asked to submit their own activity schedules to see how they fit into the overall project schedule, and to submit a list of proposed shop-drawing submittals and delivery dates of the equipment after receipt of approved drawings. When these schedules are received, the prime contractor must review them to ascertain that the equipment delivery dates fit into the overall time frame of the job. If they do, then they should be incorporated into the bar chart schedule and distributed to all subcontractors. Subcontractors should be given the complete project schedule containing their portions of work, and they should be asked to comment on the length of time and sequences allotted to them. With their input initially, and regular updated status reports, the schedule will be more accurate.

Once the prime contractor has identified the activities work items at the subcontractor microcosm level, the activities listed on the primary report file (sort by activities) are then broken out from the sort-by-activities time line and transposed onto the accelerated S-chart time line. This, in effect, shifts the production network into double-time. Superimposing the network time line S-chart on the sort-by-activities bar chart opens up details subtasks and work items within the activities, and thus, windows of float. The duration events' opening and closing begin to build a graphic fabric that should reflect the job logic. Bottlenecks appear and workarounds are developed. The overlaid events and milestones become workable and the job logic of the network schedule flows in an organized manner.

The most common cause of a schedule's failure to meet its budget is design-construction contracting in the design-lump-sum bid contract. Typically, the owner turns over too much decision-making authority by not separating the design phase from the quantity estimated takeoffs. The prime contractor at this point is not only the authority in the field work but is also in charge at the administrative level. Without shared control over all aspects of production, the natural tendency is for the contracting parties to fall into an adversarial rather than a cooperative mode. Such a scenario harms all participants, especially the owner. Owners must make an overall project schedule to check on the viability of keeping the strategic end date. They then must keep track of the overall project float to see that each major project contributor uses only that float which has been allotted to that activity. If float is shifted for any reason, the total project float must be evaluated in the new context.

Preconstruction Conference (Precon)

Certain things must occur once the best bidder has been determined to protect the optimistic environment for the project's production schedule. Prior to awarding the contract, the project team holds a preconstruction conference with all contractors. At this meeting, they:

1. Closely evaluate the job logic of subcontractors' schedules when they are presented for approval. Check their float numbers against the master schedule. Look for numbers in the time scale that don't add up.
2. Make subs prove that any overly optimistic projections are realistic within the master schedule.
3. Don't allow any prime or subcontractors to use up their allotted float early in the master schedule. Cushions and margins for error should be held back until the finish event, and preferably, not used at all.
4. Monitor their schedule reports carefully; check each report for accuracy; move quickly on any discrepancy.
5. Establish the precedence for checking actual progress in the field.

The prime contractor prepares an operational production schedule that the project manager uses for the weekly scheduling meetings held in the field office, so the operational schedule must be detailed in activity and subactivity reports to serve that objective. Operational schedules have a more detailed work breakdown structure than an owner's schedule and are better suited for accurate field reports.

During the precon, the prime contractor must ensure that all work contracts specify that ordering materials shall be done after contractors review plans and specifications. Simply visiting a job site is not sufficient to see all of the details of the job. If materials or items are allowed to be ordered on an "as-needed" basis, delays will occur if the materials or items are unavailable or on back order. The project schedule, no matter how well done, will be handicapped if this important step is not taken at the precon.

Once the contract is awarded, the project begins. Now everyone involved will be playing beat the clock with your schedule. Buckle up and be ready to ride tall in the saddle when it comes time to ramrod your schedule. From the moment the chute opens until the pay bell rings at the end, it'll be a busy rodeo.

Project Schedule Sequence

Project schedule sequence serves to set the precedence for all milestone events and their relative events and durations falling within the project's schedule. As stated earlier, and repeated here for ease of reference, these are arranged in chronological order to visually reflect the project's job logic. This type of sequential task completion listing is known as project schedule sequence. Project schedule sequence should include:

- All meetings, both home office and field
- Progress inspections, both municipal and site supervisor
- Engineering and testing requirements
- All long-lead items delivery dates and purchase orders from subs
- Equipment delivery dates
- All submittals due dates
- Prime and subcontractors progress payments schedule
- Project phases completion dates

Schedule Operations Analysis

Interlinked operations of activities and time scale systems are the basic structures of your construction project's schedule. They include everything from materials delivery scheduling to the entire scheduling process itself. The operations analysis is unique to each schedule and each project. You should review each element of your schedule in planning but should not (in the interests of time management) revise or develop a full plan for each activity and interlinked operations in every planning cycle. Look for the critical path activities and interlinked operations that are most in need of attention or offer the greatest opportunity for development of shorter events (activity durations). All schedules have the following operating systems:

- **Planning**. This is the process which you initially begin when you first start building your project schedule.
- **Organizing**. In complex projects, this is reflected in network scheduling diagrams, work-flow charts, and sorts.
- **Controlling**. Controlling is the process of creating expectations and evaluating progress effectiveness. This involves the time analyzing of sorts and making appropriate changes in the project schedule's recycling phase.

Another important area of schedule operations analysis is the determination of your network schedule's capacity and efficiency. To make cost and performance evaluations, it is necessary to know what your project schedule's current performance limits are. Capacity study is based upon an understanding of the tasks that are devoted to the activity. Good project managers continuously monitor and improve their capacity assessments. The initial assessment can be done through hypothesis or best "guess-estimate," but it is far more accurate if based on the company's historical data from similar projects. As an example, consider this simple illustration. For a single subcontractor providing services for one activity, the elements of capacity determination are the number of hours in a desirable work week (40), the average number of hours of reporting effort to produce field reports on that activity (5), and the hours per week of administrative time for the owner to produce the summary report for that activity (4).

Weekly productive capacity for that activity is 36 hours (40 less administrative hours) divided by 1.5 (1 hour of computer processing plus .5 hour of field reporting). Net capacity = 24 hours per week. This is the basis of the schedule's capacity and efficiency. The trick to matrix networking productivity improvement is to follow up with the following questions. Answering yes to any of these questions brings more scrutiny to details of the project and less ambiguity leading to change orders. How can I check the accuracy of the project's time-frame assumptions? Perhaps a daily time allocation check of estimated work done versus actual work done? Can I improve my production schedule's efficiency? Can I reduce my schedule's administrative hours in controlling the project?

Computerized Project Scheduling

Running a project schedule on computer significantly cuts down on claims and helps to quantify the impact of an owner-caused delay. Many contractors do not usually have high-degree sophisticated software in their computer programs, and the odds are against a large project network being scheduled efficiently by a contractor showing 10 different critical paths with the same duration. Invariably, this type of scenario results in unforeseen change orders and claims that typically go to litigation. This situation can often be foreseen in a contractor's network project schedule, while the same situation will not show up in project phase scheduling.

For example, the production control conditions bar might indicate a large quantity of subactivities covering workplace safety requirements, public pedestrian barricades, or specific traffic control plans. These would all be individual activities on a commercial project but are too often grouped together as one activity on residential project schedules. This type of inadequate breakout of subtasks is a problem that invariably leads to change orders. But showing the production activities and their subactivities in that sort of detail could easily take 20 pages of bar-chart schedules just to list all those phases, related construction activities, subactivities, tasks, and work items. The computer will do this in seconds, with a one-page summary report. See Figure 4-4.

The primary reason for the cost overruns and late completions of projects in the 1960s was this lack of control over the project's schedule time line, and became a focal point of change for more efficient construction management. Delay overruns on high-end residential can run into hundreds of dollars per day. Bar charts used up to that time proved inadequate. Scheduling larger projects in that sort of detail with bar charts simply does not offer the time-manipulation advantages of modem computerized project scheduling.

Computerized project scheduling makes up for the bar chart's inherent disadvantages by interlinking activities and producing integrated progress summary reports called sorts, which contain current data, available to any end user on demand. By first monitoring and recording production activities progress on each production development activity, computerized project then integrates the data through the sorts, one of which produces progress reports in a bar chart sort. Experts are saying that soon we'll have remote terminals the size of credit cards to input this data from a car back to a host computer in the office that does the work.

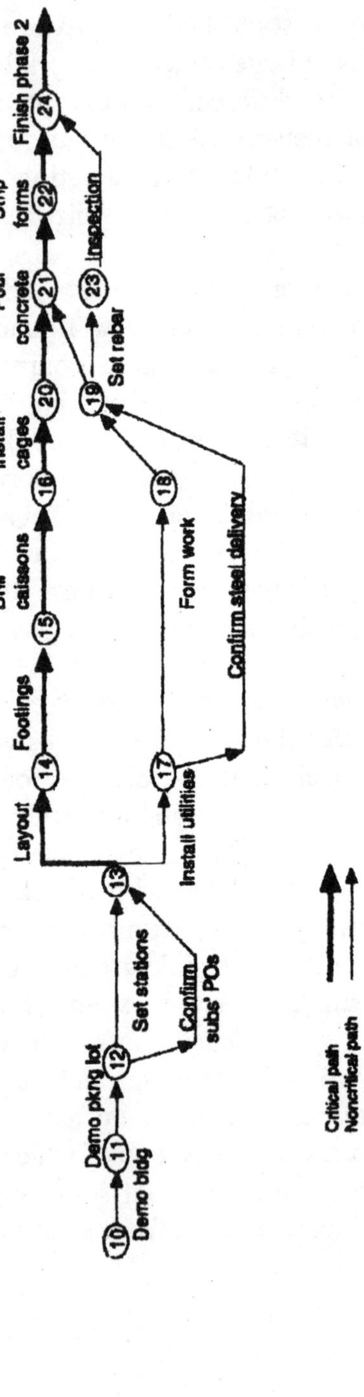

Figure 4-4
Computerized Scheduling

CPM Schedules

CPM can be thought of as the path of least float, based on the relationship between sequential activities and completion duration of the activities. Think of float as scheduling time leeway, or slack. In developing a CPM schedule, the critical path is first determined by identifying all activities with zero float time. Any activity is considered critical if its completion delay will cause total project completion delay. Critical path activities with zero float time must begin when scheduled, or the total project completion date will be pushed back. The arrow diagramming method (ADM) and the precedence diagramming method (PDM) are the two basic S-chart scheduling standards upon which CPM has evolved.

One of the cost-effective benefits achieved by using CPM scheduling is the analytical consideration of activities duration and job logic sequencing of each activity in the initial stages of the project as the CPM schedule is being developed. An example of such a CPM network diagram can be described by an arrow diagram. The tail end of the arrow indicates the start of an activity and the head end represents the completion of that activity. Using a graphic arrow to indicate paths on a flow chart, each activity (or arrow) will have a start, stop, and duration. When two or more arrows or activities meet, the intersection is called an event. Activities are begun and completed at events, and succeeding activities in their production phases move forward from one event to another event. The events are assigned numbers, which are used by the computer to change event sequences or durations as the CPM program is monitored and recycled during the production period. The various activities and events in the CPM schedule, making up the network diagram, are interlinked by interdependence and the project's time line.

A CPM schedule shows the interdependence of one activity on a preceding or succeeding activity, much like a shadow. For instance, if the excavation for the building's footings is scheduled to commence 10 days before forming activity commences and 16 days before placing of the concrete event commences, any delays due to late delivery of materials or inclement weather will cause subsequent activities to be delayed, pushing back the contractual completion date. Typically, in modern commercial projects, a structural-steel framing system will be erected on the building's foundation. So if, in our example, the earthwork excavation delayed the forming and pouring of the footings, the delay would also push back the entire phase of activities involving placement of the steel and iron workers. The CPM schedule will also show what effect inclement weather delays will have on foundation work, completion leading to the structural steel starting events and finishing events, and the relations of these changes to all other construction phases. If there is zero float time on the critical path, the project contractor using CPM can decide whether the concrete foundation work is to be accelerated or the steel erection can be delayed.

Showing interdependence of one task, or work item, on another is the major difference between bar charts and CPM scheduling methods. Whereas the standard bar chart may show a continuous line of activity for a particular trade, which starts at one point and continues uninterrupted to another point, the CPM chart draws attention to the specific starting and ending dates (events) for each major portion of a construction component. The CPM will also show the interdependence of each activity and the effect of one event finishing late or early on the starting event of the succeeding construction activity.

CPM scheduling has disadvantages, also. It will tend to increase the total contract price. Such schedules are expensive to create and maintain. It can also be subverted to build artificial scheduling networks by those who are good at using it.

This is how it's done: To prevent creation of artificial networks on the production network diagram, those schedules traversed by more than one critical path should require a contractual agreement between the owner and the contractor prior to the start of the work or any change in work's scope. To prevent manipulation of activity duration by the owner, networks with several near-critical paths should be subject to resource (labor hours, equipment hours, dollars, etc.) and quantity (permanent material) analysis with respect to activity duration. Owners and contractors should be required to review and approve such analyses prior to the start of work or start of change-of-scope work. In today's business combat, control in this area is another profit maker for the claims-conscious contractor.

As-planned, As-built, and Adjusted Schedules

Construction schedules are also used in the context of justifying time extensions and delays in project operations. The kind of analysis that can accomplish these demonstrations are either performed before the situation is actually encountered in a claim situation, or immediately afterward. In the first case, reasoning is presented to convince the owner that a potential change will affect the schedule by a precise number of days. This is accomplished with the use of two separate but related schedules. The first is the as-planned schedule. This is the schedule that indicates the new sequence of work, incorporating all the influences of the change along with the ultimate effect on the project end date. That end date change is attributable to the change.

The second situation is the one in which construction schedules must be used to demonstrate where you would have been had it not been for the subject change. It involves an "*ex post facto*" analysis, often in a delay claim situation. This particular kind of analysis is most useful in those cases where the project has been affected by a change, but the usually complex interaction that makes up the total project history confuses any clear cause-effect relationships. The dilemma is solved with the use of three schedules. First, again, is the as-planned schedule. Next is the historical as-built schedule that through its periodic updates incorporates all the complicated effects on the construction sequence.

Finally is the adjusted schedule. This schedule reconstructs each update as necessary, removing the effects of the change of interest. The result is a schedule that includes all the other effects without the effects of the change being considered. The difference in the end dates between the as-built and adjusted schedules becomes the amount of time attributable to the single change. The conditions bar might have a large quantity of subactivities covering workplace safety requirements, public pedestrian barricades, or specific traffic control plans. These would all be individual activities in the adjusted schedule on a larger project. But showing the production activities and their subactivities in that sort of detail could easily take 20 pages of bar-chart schedules just to list all those phases related construction activities, subactivities, tasks, and work items.

The primary reason for the cost overruns and late completions of projects in the 1960s was this lack of control over the project's schedule time line, and became a focal point of change for more efficient construction management. Bar charts used up to that time proved inadequate. Delay overruns on high-end residential projects can run into thousands of dollars per day. Scheduling larger projects in that sort of detail, with bar charts, simply does not offer the advantage of reducing change orders like modem computerized project scheduling. Computerized scheduling makes up for bar charts' inherent disadvantages by interlinking activities and producing integrated progress summary reports called sorts, which contain current data, available to any end user on demand. By first monitoring and recording production activities' progress on each production development activity, computerized project scheduling then integrates the data through the sorts, one of which produces progress reports in a bar-chart sort. Here the situations leading to change orders can be identified.

PERT Project Schedules

The PERT project-scheduling system was developed in the military for government production scheduling control while contracting with multiple and different types of defense contractors. PERT (program evaluation review technique) is a project-scheduling system that uses inside and outside figures to make a best-guess estimate on each activity duration. These "guesstimates" are then strung together by the prevailing job logic. In the PERT system, the elapsed time of an activity is calculated by assessing an optimistic activity finishing event date and a pessimistic activity finishing event date, then calculating an average duration of the activity. This average time is then used as the only factoring used for activities durations in the network schedule. This "inside and outside best-guess estimate" will work for averages in data processing, but PERT does not account for potential change orders.

In the PERT system of project scheduling, generalized time estimates for activities throughout the total project are made, with timing leeway allowed for optimism or pessimism of activities duration. However, the linkage of activities and establishing a critical path through a PERT system is very difficult and it is impossible to maintain workaround options in the event of changing critical paths. To be accurate in event time estimating, the contractor breaks out all activities into manageable units of time. The elapsed time estimate is based on the unit labor and materials quantities taken off the project's specifications, which are used to estimate the project cost. This quantitative survey estimate contains a wide variety of bidding pre-estimated information, such as cubic yards of import fill or excavation, tons of structural steel, numbers of block masonry units, and cubic yards of concrete.

That information, combined with previously recorded labor and material costs from the company's own records or activity subcontractor, is used to calculate the expected elapsed time duration for each activity. The unit number is used as a computation factor and is then adjusted for delay by factoring in prevailing local conditions, such as seasonal inclement weather, variations in delivered materials costs, area union labor rates, trucking strikes, etc.

Accordingly, scheduling programs for residential construction projects do not include the PERT-type scheduling system, but instead use the data from precise activity events, both early and late, calculate with all four of those factors, then average the duration. Float can also be taken into account to offset change orders. The proper selection of the elapsed duration times for the work activities is an important part in preparing an accurate network schedule because exact events determine the exact activity duration times, which in turn directly affect the critical path and succeeding activities float times. The PERT system does not offer these scheduling advantages.

Scheduling Philosophy

The dichotomy between scheduling planning and real-world construction execution are the two main factors affecting the contractor's preparation of the project schedule. Practical limitations of the project's production must be established, then prepared for in the planning stages. We call this conception and critical thinking stage *scheduling philosophy*. Once the project commencement occurs, the basic ground rules for its construction schedule must have been already established. These will address the scope of construction phases, materials procurement procedures, contingency critical paths planning, and long-lead items procurement. These ground rules will be different in specifics with the job logic of each project, but large or small, they will all be determined by following basic scheduling philosophy. Scheduling philosophy refers to the selection of the scheduling system. The questions that must be answered prior to the first scheduling meeting will determine the scheduling philosophy, which in turn will determine the production contracting basis. Scheduling philosophy uses critical analysis to determine the identification of critical path activities, workarounds (alternative option activities that can be substituted if a critical path activity hits a bottleneck), job logic (sequence of activities), timing of schedule recycling, and the contractual requirements for progress reporting from field subcontractors.

Typically, it is the owner's responsibility to establish the scheduling philosophy, critically analyze decisions, then get owner approval of scheduling system selection. The scheduling philosophy must be a cohesive decision among all parties because changing the system later during the project's phases is disruptive of the production continuity and will result in costly change orders. The scheduling philosophy should also account for the start-up sequence of the various activities making up the production phases. Activities that start up first must also be the activities that finish first. The project schedule must also interlink with the other major project considerations, such as design and long-lead items procurement. That is easy enough where, traditionally, the design is completed before the project goes to construction bidding. But in other production execution modes, such as fast-tracking, design should be 30% to 50% complete if a reliable project schedule is to be developed and executed. By getting this advance start on a project without a finished projects design, change orders are probable if the operation has not been done previously to provide accurate activities duration.

Field Schedules

Even a detailed project schedule with a comprehensive work breakdown structure is not an adequate tool for controlling the daily progress of the activities in the field. Field scheduling is necessary to coordinate reporting data with the main scheduling system. On smaller construction projects, the production progress milestone schedule may be the only one used. Sequential completion of phases is simple enough on small jobs and usually will be adequate. On larger projects, however, the project schedule also functions as the basis for making more detailed weekly work plans in the field, for each major activity. In either case, the approved project schedule will, in turn, determine the type and detail of the field scheduling used by the owner and the prime contractor to set the production phases milestones and monitor the projected schedule progress versus the actual completed progress. Detailed field scheduling is necessary continually throughout the entire project production to interlink the two for critical analysis and control.

Complete detailed field planning makes the most efficient use of field manpower on the priority list of tasks within those activities that are required to meet the schedule's milestone dates. This detailed planning first concentrates on those items of work that are on the critical path, then on secondary activities that are noncritical. The contractor is responsible to the owner for planning the construction activities each week for the following week. Field schedules list out the work activities sequentially that are scheduled to be completed by the following week. This list of pending activities is then discussed in the weekly scheduling meeting, with the owner.

Prior to the meeting, the contractor will have checked to see that all the necessary labor, materials, and equipment are on site to perform next week's scheduled work. The contractor will then report production readiness or deficiencies at the meeting. After the meeting, the weekly task lists are distributed to all responsible personnel, so the activities subcontractors can plan weekly production execution. Scheduled work not completed in the previous week is also examined at the meeting, and plans are made to bring those activities up to speed to avoid change order generation. The outlook for the next two weeks is critically analyzed and projected. The contractor then prepares the longer-range quarterly field scheduling plans by checking the critical milestone dates four weeks ahead, and plotting the appropriate course(s) for the next quarter.

Proposal Schedules

The project schedule first comes under serious inspection by the contractors when the bidding documents or the request for proposal (RFP) arrives at the individual contractor's office. The RFP or bidding documents contain a section devoted to the construction schedule. Usually the documents ask the contractors to develop preliminary bar chart schedules showing the major construction activities (and their durations) in the contractors' proposals. The purpose of that request is to get the contractors to look at their strategic end dates in general terms to see if they are still feasible. The owner uses the contractors' preliminary bidding schedule to compare it with the schedule the architect has developed. This is typically the architect's first opportunity to get the activities subcontractors' professional opinion as to how long the activities' construction work will actually take. This is then compared to projected scheduled duration and adjustments made accordingly.

Now, the subcontractors are not paid for doing this; it's just part of the business. So typically, subcontractors don't like to spend much time developing construction schedules during the bidding process. At this point they are not assured of getting any contracts. The only exception to this reluctance of subcontractor input is when the selection of that activity subcontractor may depend on a previous track record of being able to meet a tight completion date. This is normally the case in fast-tracking project execution. Then the subcontractor's schedule becomes a bidding edge that is worth the extra bidding expense to the subcontractor. In the case of fast-tracking projects, the subcontractor's proposed schedule will undergo intensive examination by the owner and architect during bid openings and must be viable within the owner's project schedule to avoid generating change orders.

At this proposal stage, the subcontractor must evaluate the schedule in relation to the contracting plan and the construction technology available to improve the production time line. For example, there could be contract terms and conditions affecting the contracting plan that could be changed for this project to improve this particular project schedule. If there are, the new suggestions are then reviewed by both the owner and the prime contractor to determine if the proposed changes are viable. This is the same procedure for substituting construction technology improvements that will shorten an activity's completion duration. As a professional contractor, you must be aware of where the "hot spots" are that will typically generate change orders. Proposal scheduling is a critical juncture of one of the hotter spots. Here you benefit from your experience and critical analysis.

Scheduling System Selection

The selection of the scheduling system used in running the project's production is understandably crucial to either preventing or generating change orders. To be both efficient and cost-effective, the selection must consider the size and complexity of the project, scope of services required, the sophistication of users (owner, project team, field personnel, subcontractors, etc.), available scheduling programs for the system, and owner preference.

Your own experience with present systems is an important factor, because using a system with which you are familiar, that has proved itself reliable on similar projects, improves manpower time-efficiency in operating the system. Introducing a new system on a project already in progress causes more problems than it solves and is often the kiss of death. Accordingly, the curve of cost-effective schedule control usually goes down upon initial introduction and use of a new scheduling system, until the system becomes comfortable and familiar to those who use it. It follows, then, that the time for training personnel in the use of any new system is before operations commence.

The minimum hardware requirements for computerized project scheduling are an IBM-compatible 486 PC of 640K or better memory, an 80-megabyte hard disk or better, and a laser printer. Presentation graphics require a plotter, but that is not essential to doing a decent job of schedule control. The cost of the software program training is the hidden factor in the scheduling cost budget, and this depends on the experience and computer literacy of the contractor. But regardless of the cost, training is crucial to the success of computerized scheduling systems implementation. If a scheduling system is not run by trained personnel, it will not be cost-efficient. The cost-effective savings in using a computerized schedule comes from the generation of much more relevant and usable data than is possible with bar charts. Data generated by computer has a lower unit cost than labor-intensive, hand-generated data. However, again, if that computerized data is being processed by untrained personnel, the system is not being used to its fullest advantage and will not prove to be cost-effective in the long run.

Integrated Scheduling Systems

Computerized project management consists of integrated activities with logic based scheduling (see Figure 4-5) . These systems include durations, planning, scheduling, and performance-measurement aspects. These controls have to encompass the basic business parameters of progress payments and time billing, computer assisted design (CAD), cost and time estimating, cost databases, job cost, equipment and inventory costing, procurement, contract document control, and administration. All these facets must be integrated to "wash through" automatically to the primary levels of the project's accounting, which are AP (accounts payable), AR (accounts receivable), and PYRL (payroll). Even though your project schedule is only one factor in the entire project process, you can see how important it is and how it will influence identifying potential change orders. So just being aware is not enough. You must understand the link between the project schedule and costing areas to see the big picture and prospect for change orders. A fully integrated scheduling system will have this advantage.

The results of these "what-if" calculations (or other performance calculations) should be agreed upon by all parties to each activity under potential dispute prior to performing the work. To prevent improper consumption of float by either party, distributed float calculations can be used. Distributed float allocates float to each and every activity. Using such a targeted approach will always predetermine who gets what float. Relative to which way you want to go, you now know how to swing the scales on your behalf.

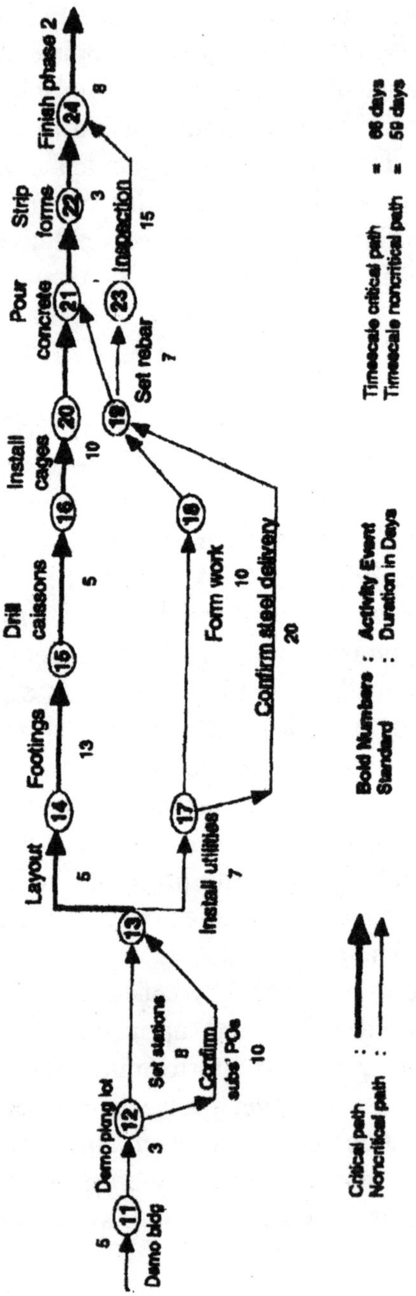

Figure 4-5
Logic-Based Scheduling

Delay Claims

AIA Document A201, Section 4.3.1, delineates the parameters of acceptable and enforceable delay claims submitted by the contractor. These requirements can make it difficult for a contractor who does not keep detailed weather records to claim a time extension for adverse weather conditions. They reflect a belief by the AIA that weather generally is a risk assumed by the contractor, and that only in extraordinary circumstances should weather be the basis for a time extension. The legal documents used by the courts to determine judgments in weather related claims are a Daily Inspection Report (DIR) and a Quality Assurance Report (QAR). If the owner is unaware of these documents, then certainly you, the contractor, should be. And you should implement them on your behalf to protect your schedule.

Another reason for not granting a time extension is the single contract system's objective of centralizing administration and responsibility in the prime contractor. Only if subcontractor-caused delay is specifically included should it excuse the prime contractor. This needs to be stated within the contract specifications to protect your client from the schedule's default through circumstances beyond your control. The independent contractor rule, though subject to many exceptions, relieves the employer of an independent contractor for the losses wrongfully caused by the latter.

Prime contractors sometimes assert that the subcontractor is an independent contractor, inasmuch as the subcontractor is usually an independent business entity and can, to a large extent, control the details of how the work is performed. Even so, the independent contractor rule does not relieve the employer of an independent contractor when an independent contractor has been hired to perform a contract obligation, and the party who suffers the loss caused by the independent contractor is the person to whom the contract obligation is owed. In the construction contract context, the owner usually permits the prime contractor to perform obligations through subcontractors.

This does not usually mean that the prime contractor is relieved of its obligation to the owner, unless the owner specifically agrees to exonerate the prime contractor. The residuary power of agency granted to the architect to grant time extensions has been criticized as leading to a deterioration of any fixed completion date. It does have the virtue of not forcing the architect to think of every possible event and include it in the contract specifications as a "catalog of events" justifying a time extension.

Another portion of contract law affects the project schedule, and that is AIA Doc A201, Sections 4.3.3 and 4.3.8.1, which cover time extensions in construction contracts. These usually provide a mechanism under which the contractor will receive a time extension if it is delayed by the owner, or by designated events such as those described in the contract specifications. Increasingly, contractors make large claims for delay damages. As a result, it is becoming even more common for clauses in the contracts, particularly public contracts and private ones drafted with the interests of the owner in mind, to attempt to make the contractor assume the risk of owner-caused delay.

An indirect technique for limiting delay claims is to require written notice by the contractor, to the owner or the project team, if events have occurred that later will be asserted as justifying delay damages. This is the method that is currently taking preference nationwide in the construction industry because it warns of potential claims in time to resolve the situation at the activity stage. Typically, this notice is stated to be a condition precedent to any right to delay damages. Historically, although such clauses have not been looked on favorably by the court, noncompliance with a notice provision can be the basis for barring a claim for delay damages and, as a contractual obligation, can "bulletproof" the situation as well as is currently possible without undue and unjustifiable constraint on the contractor.

No-Damage Claims

Looking at the troublesome questions of the validity of no-damage clauses is not within the parameters of the contractor but lies instead with the owner. You need to be aware of alternatives so that you will have viable options to propose if you should so wish. One method used to eliminate or reduce the likelihood of such claims and delays to your schedule is to have the contract specify that the owner does not have the right to delay the contractor, and that such interference is a contract breach.

Another common method to deny the delay damage problem outright by not setting up notice conditions or specifying what can be recovered is to configure the contract with no-damage or no-pay-for-delay clauses. Such clauses attempt to place the entire risk for no-damage damages on the contractor and to limit the contractor to time extensions. Generally, such clauses are upheld but are not looked on favorably by the courts and much less by contractors. A current modification on the no-pay or no-damage provision clauses is to provide, that the contractor can recover delay damages after a designated number of days delay by the owner.

A newer technique is to have contract clauses that permit some delay damages but avoid large claims for diminished productivity by an open-ended, total-cost formula. The method provides that the contractor can recover delay damages, but only for premiums paid on its bond and for wages and salaries of workers needed to maintain the work, the project materials, and the equipment during the delay. California courts, however, have held that such clauses will not be enforced if there is affirmative or positive interference or a failure to act in some manner essential to the prosecution of the work.

Whether such clauses should be included in a contract depends both on the likelihood of their being enforced in a proper case, and on the willingness of the owner to take "front-loaded" costs (at the front end of the contract) rather than through claims at the end of the work. However, the use of these clauses in a highly competitive construction market will protect the project schedule somewhat by encouraging bidders to bid low and take their chances.

Project Schedule Manipulation

Network diagram scheduling systems were originally developed as production management and cost-control tools in Europe, but like everything else efficient at making money, network diagram scheduling has evolved significantly. Owners learned how the system could be abused for their own advantage. Some developers and project teams ha\e even used CPM scheduling as a litigation weapon to secure as much profit as possible through denial of claims by contractors.

Conversely, prime contractors and subcontractors have been known to employ professional scheduling tactics to devise ways to give their particular business an edge on float in activity events, especially those on critical path(s). Modem computer network-based management systems that run a construction project like a mathematical clock can also create biased documents to support the originator's right to claims and extra work orders. Those who are familiar with software spreadsheets will nod in agreement with me here as I inform you that anyone can make wrong data look right in charts if it's presented with enough spreadsheet manipulation and corresponding bar graphs.

On one hand, a contractor might create an artificial network with multiple critical paths. The intent of the contractor would be to present claims if the owner caused delay on any of the paths. On the other hand, the owner might plan the project duration and then shorten it. The owner's intent would be to obtain a bid on the shortened duration then hold the contractor to the time limit. Obviously, these practices reduce an otherwise effective management tool to a weapon for justifying or denying claims. The result has been more litigation over more claims. See Figure 4-6.

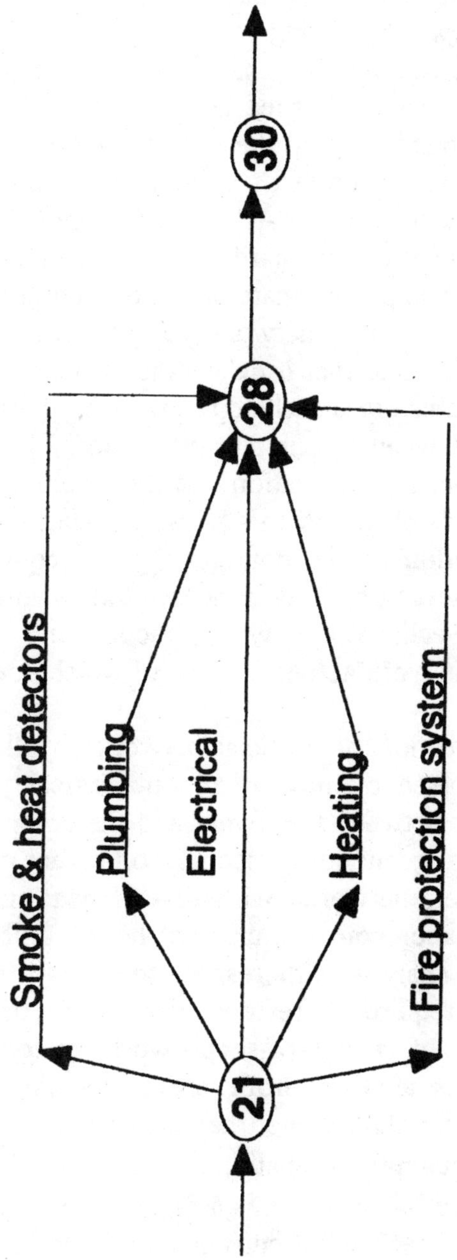

Figure 4-6
Parallel Activities

Contractors and owners view the problem differently. A contractor might question whether an owner has a right to direct starts or delays for specific purposes. Owners, trying to manage project costs and overall scheduling, might ask whether they have the right to force the contractor to start an activity before an established late-start date. Too many of these late starts could cause problems toward the end of the job. For the owner, a week's slippage in a multiresidential project's completion date translates into a loss of several hundred thousand dollars. At the same time, improper or premature activity starts can cause serious hardship for a contractor, who may incur additional expense for equipment, material, labor, and other resources.

Very little float is in a network velocity diagram, and the owner is subject to claims at any time that it delays the prime contractor on any one of the multiple critical paths. This technique invariably gets the prime contractor off on the wrong foot when he submits the claims. The owner's project team will then resent the owner's implication that they were not smart enough to see through the contractor's ploy, and an adversarial relationship will be set up that will last for the duration of the contract. When requiring a CPM schedule submittal from a contractor, it is wise to follow the Associated General Contractors guide specifications, which recommend that the network be comprised of individual work activities, each of which does not exceed 10 days' duration.

To prevent creation of artificial networks by the designer or owner, those schedules traversed by more than one critical path should require a contractual agreement between the owner and the contractor prior to the start of the work or any change in work's scope. To prevent manipulation of activity duration by the owner, networks with several near-critical paths should be subject to resource (labor hours, equipment hours, dollars, etc.) and quantity (permanent material) analysis with respect to activity duration. Owners and contractors should be required to review and approve such analyses prior to the start of work or start of change-of-scope work. In today's business combat, control in this area is another profit maker for the hot contractor. Such performance analyses should consider the following:

- Budgeted rate of quantity installation as a function of time.
- Budgeted rate of resource usage as a function of time. This can be done by resource-scheduling techniques currently available in most CPM processors.
- Budgeted consumption of resource usage as a function of quantity, type, and time.

The results of these calculations (or other performance calculations) should be agreed upon by all parties to each activity under potential dispute prior to performing the work. To prevent improper consumption of float by either party, distributed float calculations can be used. Distributed float allocates float to each and every activity. Using such a targeted approach will always predetermine who gets what float. You now know how to swing the scales on your behalf. However, remember, information is power. These tactics are the domain of the professional contractor and professional trade secrets should not be confided to others on less than a need-to-know basis. As Shakespeare said, "Once the deed is done, 'tis done forever and cannot be undone." Your prowess in these matters may or may not be something you wish others to know about. It's therefore recommended that your knowledge of manipulation techniques in project scheduling and network diagramming be kept close to your vest.

Time Extensions

Another portion of contact law affects the project schedule, and that is AIA Doc A201, Sections 4.3.3 and 4.3.8.1, which cover time extensions in construction contracts. These usually provide a mechanism under which the contractor will receive a time extension if the project is delayed by the owner or by designated events such as those described in the contract specifications. Increasingly, contractors make large claims for delay damages. As a result, it is becoming even more common for clauses in public works contracts to attempt to make the contractor assume the risk of owner-caused delay. Owners are limited by time constraints and funding issues, so they must contract in advance for the full cost of the project. To do this, many owners use the disclaimer system for unforeseen subsurface conditions. Similarly, they wish to avoid facing claims at the end of the project, based on allegations that they have delayed completion or required the contractor to perform its work out of sequence. There are two basic management techniques for limiting delay claims in a project schedule:

The first management technique is to require written notice by the contractor to the owner or project team if any events have occurred that later will be asserted as justifying delay damages. This is the method that is currently taking preference in the industry, as it provides warning of potential claims in time for someone to be able to do something about the situation in the mediation-dispute-resolution stage, and resolve it at the activity level. Typically, this notice is stated to be a condition precedent to any right to delay damages. Historically, although such clauses have not been looked on favorably by the courts, noncompliance with a notice provision can be the basis for barring a claim for delay damages, and because it's a contractual obligation, can "bulletproof" the situation as well as is currently possible without undue and unjustifiable constraint on the contractor.

The second management technique is to deny the delay damage problem outright by not setting up notice conditions or specifying what can be recovered. The method is to configure the contract with "no-damage" or "no-pay-for-delay" clauses. Such clauses attempt to place the entire risk for delay damages on the contractor, and to limit the contractor to time extensions. Generally, such clauses are upheld but not looked on favorably by the courts and much less by contractors. A current modification on the "no-pay" or "no-damage" provision clauses is to provide that the contractor can recover delay damages only after a designated number of days' delay by the owner.

AIA Document A201, Section 4.3.1, delineates the parameters of acceptable and enforceable delay claims submitted by the contractor. These requirements can make it difficult for a contractor who does not keep detailed weather records to claim a time extension for adverse weather conditions. They reflect a belief by the AIA that weather generally is a risk assumed by the contractor, and that only in extraordinary circumstances should weather be the basis for a time extension. The legal documents used by the courts to determine judgments in weather-related claims are a Daily Inspection Report (DIR) and a Quality Assurance Report (QAR). If the contractor or owner is unaware of these documents, then certainly you, the professional project contractor, should be. And you should implement them on your client's behalf, whomever that may be, to protect your schedule.

Another reason for not granting a time extension is the single contract system's objective of centralizing administration and responsibility in the prime contractor. Only if subcontractor-caused delay is specifically included should it excuse the prime contractor. This needs to be stated within the contract specifications to protect your client from the schedule's default through circumstances beyond your control. The independent contractor rule, though subject to many exceptions, relieves the employer of an independent contractor for the losses wrongfully caused by the latter.

Prime contractors sometimes assert that the subcontractor is an independent contractor, inasmuch as the subcontractor is usually an independent business entity and can, to a large extent, control the details of how the work is performed. Even so, the independent contractor rule does not relieve the employer of an independent contractor when an independent contractor has been hired to perform a contract obligation, and the party who suffers the loss caused by the independent contractor is the person to whom the contract obligation is owed. In the construction contract context, the owner usually permits the prime contractor to perform obligations through subcontractors. This does not usually mean that the prime contractor is relieved of its obligation to the owner, unless the owner specifically agrees to exonerate the prime contractor. These are known as constraints (see Figure 4-7). The residuary power of agency granted to the architect to grant time extensions has been criticized as leading to a deterioration of any fixed completion date. It does have the virtue of not forcing the architect to think of every possible event and include it in the contract specifications as a "catalog of events" justifying a time extension.

Another portion of contract law affects the project schedule, and that is AIA Doc A201, Sections 4.3.3 and 4.3.8.1, which cover time extensions in construction contracts. These usually provide a mechanism under which the contractor will receive a time extension if it is delayed by the owner, or by designated events such as those described in the contract specifications. Increasingly, contractors make large claims for delay damages. As a result, it is becoming even more common for clauses in the contracts, particularly public contracts and private ones drafted with the interests of the owner in mind, to attempt to make the contractor assume the risk of owner-caused delay.

Public entities are often limited by appropriations and bond issues. As a result, they must know in advance the ultimate cost of a construction project. Thus, many public entities use the disclaimer system for unforeseen subsurface conditions. Similarly, they wish to avoid having claims made at the end of the project, based on allegations that they have delayed completion or required the contractor to perform its work out of sequence. These public works and awarding-authority entities recognize that barring claims may cause higher bids, but they would prefer to see bidders increase their bids to take this risk into account rather than to face claims at the end of the job.

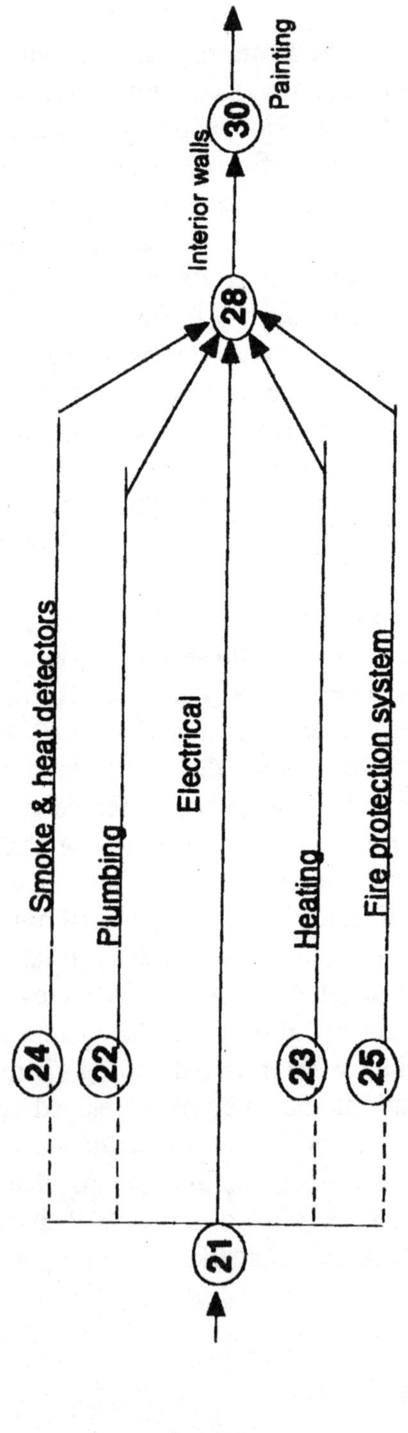

Figure 4-7
Constraints

Chapter 5

Contract Change Orders

Change Orders

Contract law does not allow the prime or subcontractors to make any changes to an existing contract. Any addendum in specified work requires the contractor to write up a contract change order and then seek written approval from the owner. The owner is not obligated to sign a change order, but the owner's approval is absolutely necessary to avoid litigation problems. The owner cannot be forced to pay for work that was not authorized. Contract law specifies that "all changes to the original contract must be in writing and signed by both the property owner and contractor." Most construction contracts carry general wording that specifies materials ordering shall be done after the contractor reviews plans and specifications. Simply visiting a job site is not sufficient to see all of the details of the job. During project operations, if materials or items are ordered on an "as needed" basis, delays will occur if the materials or items are unavailable or on back order.

Contract Change Orders

A contract change order is an authorization of work in addition to, but pertaining to, the original contract. The courts have typically held that a change order is a contract in itself, and since it amends the original contract, it must be reflected in the network schedule and filed with the original contract documents. The legal sequence that must be observed regarding change orders is that, if the network schedule requires numerous changes during the course of the project, the contractors must bill the owner after the changes have been completed. This is important to know, because to do otherwise is to allow your project schedule to become sidetracked or derailed. If you have to play hardball with the subcontractors to keep your schedule on track, so be it. The law says that a subcontractor may collect (in progress payments) only what he has provided in labor and materials to date. If, on Monday, the contractor performs additional work that amounts to $5,000, the contractor can amend (through a contract change order) the next progress payment to include that amount. However, the work must be done first.

Contractors have a right to collect extras as long as they amend their contract and/or use change orders indicating when extras are being added and when they are to be paid for. This means that activities cannot be suspended by the contractors for extra work payments, and that your schedule cannot be brought to a halt over contract change orders. The primary objective of any project schedule is to complete the project as designed, in a systematic, coordinated manner. All this must be accomplished in the shortest time consistent with material and personnel constraints, thereby maintaining a good profit.

Today's construction industry management professionals recognize that change orders are a normal part of the construction process. Every project manager, site superintendent, and project developer knows that, in all probability, there will be changes occurring on the project that will affect the final duration. If they can be handled competently and in stride, the effects on the project might still be felt, but accountability for them will remain where it belongs.

Project schedules that are constructed and updated in a clear, consistent manner with complete, thorough notes and references can accomplish this. If the schedules are managed so that the information in them can be easily extrapolated, categorized, and supported (or denied) in the event of a worst case scenario, their strength will actually minimize serious disputes. A good schedule will easily provide significant, visible comparisons and convincing proof of damages, including delay, acceleration, suspension of work, inefficiencies, disruption, interference, and demobilization.

Billings and Payments

Change order billings and payment will normally follow the same procedure and timing as that for contract work, unless specific conditions are negotiated as part of the final proposal. The attention given to any special billing circumstances will depend upon a variety of factors. These include any money required up front to finance the change (such as a payment to a utility company before their work is even scheduled); the size of the change order, and its corresponding retainage amounts; delay effects on the ability to invoice for contract work; and delay in the date of substantial completion or other significant date that determines when the retainage on the entire contract amount can either be reduced or eliminated.

There can be any number of other considerations, depending upon the contract itself or even the current billing format. It is common for the contract Schedule of Values to include an item for General Conditions or something similar. If so, it may be billed each month in the same percentage as the total percentage complete for the project. If that's the case, a very large change order that has just been added to the books but that has not yet been completed will have the arithmetic effect of lowering the total percentage complete for the entire project, given the new increase in total project cost. For the same amount of physical work accomplished to date, the General Conditions item will actually be billed at a reduced rate. It is important for the builder to forecast these kinds of effects and accommodate them in the most reasonable way in the change order proposal's first submission.

Constructive Changes

A constructive change is an informal act authorizing or directing a modification to the contract, caused by the owner or architect through an act or a failure to act. In contrast to the mutually recognized need for a change, certain acts or failures to act by the owner that increase the builder's costs and/or time of performance, may be considered grounds for a contract change order. This is termed a *constructive change*. However, it must be claimed in writing by the builder within the time specified in the contract documents. Otherwise, the builder may waive its rights to collect. Types of constructive changes include:

- Defective plans and specifications
- Architect or engineer interpretation of documents
- Higher standard of performance than specified
- Improper inspection and rejection of work
- Change in the method of performance by owner
- Change in the production sequence by owner
- Owner nondisclosure of pertinent facts
- Impracticability or impossibility of performance

Sometimes referred to as "unforeseen conditions" or "differing site conditions", the term *changed conditions* is typically used in most construction contracts. There is also a growing trend by many public agencies and a few private owners to adopt similar wording in their contracts. Failure of an owner to provide payment for changed conditions places the builder in a difficult position. If the owner takes a hard-line position on this issue, the builder may find it necessary to seek relief from the court, in a claims process that is both lengthy and costly to both parties.

A constructive change order can arise not through the stated change procedure, but through the conduct of the parties. Actions or inactions by authorized owner representatives may change the scope of the work and increase the cost of performance. If so, the contractor would be entitled to an "equitable adjustment" of the contract amount. Any actions by the owner requiring responses from the contractor that are different from those necessary by the terms of the original contract may constitute a constructive change order. A constructive change may be said to have occurred in cases where:

- Extra work is known by the owner to be required and not objected to.
- Extra work is ordered by the owner or its representatives either orally or in writing.
- An approved change results in other additional work not anticipated by the original change order.
- The parties consistently disregard the requirement for written change orders as a regular course of doing business.

The key to effective resolution and payment for the constructive change order is immediate recognition of it for what it is. The contractor who simply proceeds with the work of a constructive change without prior confirmation of the condition does so at great risk. The Change Clause entitles the owner to order changes in the work. As such, there is also a risk if a contractor refuses to perform the work of a constructive change. If written confirmation of the change cannot be secured from the owner by the time the work must be accomplished, it may be effective to notify the owner that you are proceeding with the constructive change under protest. If the item is significant, however, any of a number of actions may be appropriate. It may be smart at that point to get your attorney on the scene.

The court's policy is to make adjustments in time and/or price where unknown subsurface or latent conditions at the project site are encountered by the builder. The purpose is to have the owner accept certain risks and thus reduce the large contingency amounts in bids to cover such unknown conditions. The AIA (American Institute of Architects) and EJCDC [Engineers Joint Contract Documents Committee (see Appendix)], recognized nationwide as the authoritative industry standards for construction contracts as well as other document agencies recommend provisions in their construction contracts that will grant a price increase and/or time extension to a builder who has encountered subsurface or latent conditions.

Under the legal definition of changed conditions, an existing underground pipeline that was either not shown on the plans at all or was incorrectly located on the contract drawings would qualify as a changed condition. Unusually severe weather conditions for the time of year and location of the project may also qualify. The discovery of expansive clays in the excavation areas, if not accounted for on the geological site survey and not detected in prior soil investigations, also qualify as changed conditions. Severe rains or similar weather that prevent work from being done, or which in any way delay the project, may not always be excusable delays, and in some cases have been ruled by the court to be excusable only and not compensable.

Any scheduling change will have an impact on total project operations and can even push back the contractual completion date. When scheduling changes are implemented as a result of constraints created by the owner or architect, the basis for a builder claim may exist. Whenever the owner or architect issues any change order, field order, or work directive, or whenever a constructive change condition exists in any of the following areas, the builder will begin to consider the impact of such action by the owner or architect upon his project costs and profitability. These are the typical schedule contingencies that need advance alternative planning. They can lead to schedule changes that can later develop into owner-or architect-caused delays:

- Requirements to deviate from the schedule
- Orders to expedite the work
- Job interference by the owner or architect
- Owner constraints on scheduling interfaces
- Late availability of owner-furnished materials
- Owner-imposed acceleration or deceleration
- Impractical or impossible milestones
- Extra work orders
- Contract change orders
- Schedule impacts

- Sequence of construction altered by owner or architect
- Changes in activity's start event
- Changes in activity's finish event
- Time extensions
- Utilization of scheduling float time
- Late schedule approvals

If a CPM project schedule is not available during the production, any consistent scheduling method that illustrates the delay will help support the claim. Any change in the project schedule, imposed by either the owner or the architect, can become a basis for a claim by the prime builder or activity subcontractor. The best way to prevent a claim is to utilize CPM scheduling techniques. This form of scheduling documentation offers the greatest protection. As the project progresses, time extensions may be granted by the owner, or by the architect with owner approval. These would result in changes in the project schedule and could result in revised interface dates or changed completion dates.

The builder needs to keep a close watch on the status of the schedules that it submits to the owner or architect for approval. Departures or deviations in the builder's schedule from the schedule originally intended by the owner may be considered as contract amendments after approval by the owner. If a builder submits a shorter schedule than originally called for under the contract, and it is approved by the owner or the architect, any liquidated damages provided under the contract may now be applied to the early finish date shown in the shortened schedule. There is no question that the only way for a builder to make a profit on a project is to get in and get out in the shortest possible time, thus reducing the office overhead and labor costs allocated to that job.

Contract Documents Correlation

The mechanism to decide between certain types of conflicts in the contract documents may or may not be completely detailed in the specifications. If attention is given to the issue, its comprehensiveness is likewise varied. Some specifications will simply state that in instances of conflicting details or descriptions, "the more expensive will apply." Other specifications are more responsible, outlining the precise rules of precedence. For example, addenda take precedence over the specifications, specifications take precedence over the plans, schedules take precedence over other information in the plans, large details take precedence over small details, and stated dimensions take precedence over scaled dimensions.

While it is true that the foregoing is little more than a restatement of the rules of common sense, they are very helpful in dealing with otherwise senseless directives. Know if such descriptions exist in your contract and use them when necessary.

Cost Perceptions

There are always two value systems at work in any negotiation. The value of extra work to you is closer to your costs. The value of the work to the owner may extend far beyond the addition of your overhead and profit. The work in question may have other bearings on the owner's ability to continue with its own affairs. For example, if you fail to reach agreement on a change, the cost to you may be the cost of the work. The cost of the change to the owner, however, may be the cost plus contractor overhead and profit, plus the opportunity cost resulting from the inability to use the project by a certain point in time. Be observant to the perceptions of the total cost (real and potential) for a change. You'll then be in a better position to assess the strength of your hand accurately in the negotiations.

Contract Acceleration

Acceleration of the contract work is usually the result of an attempt by the owner to apply whatever means and take whatever measures are necessary to complete the work sooner than would normally be expected for a given project under stated contract conditions, or an attempt by the builder to take extra measures to make up for delays, whatever the cause is, by utilizing whatever means are at the builder's disposal to accomplish such an objective.

There are two recognized types of schedule acceleration:

- Directed acceleration
- Constructive acceleration

A directed acceleration occurs when the owner or architect, at the owner's direction and authority, orders a builder to speed up the work. The owner is definitely in this legal position if it advanced the builder's finish date. The necessary elements of a constructive acceleration claim have been outlined by the U. S. District Court as follows:

"Constructive acceleration is present when

1. The builder encountered an excusable delay entitling it to a time extension;
2. The builder requested an extension of duration time;
3. The request was refused;
4. The builder was ordered to accelerate its work, that is, to finish the project as scheduled despite the excused delays; and
5. The builder actually accelerated the work."

In considering the liability of acceleration, the court's first consideration will be whether the so-called "acceleration" is the result of an order from the owner, or its agent, to a builder who is behind schedule to get back onto schedule, or an order from the owner, or its agent, to the builder requiring that the builder, either directly or constructively, to complete the work prior to the scheduled completion date.

If the builder is behind schedule and it is the desire of the owner, or its agent, to require the builder to get back on schedule, it is sometimes necessary to direct the builder to take whatever means are necessary to assure completion by the originally scheduled completion date. Herein lies an administrative risk for the builder and contract administrator. If an order is issued to the builder directing it to accelerate the work so as to catch up or make up for delays or lost time, the risk is that the builder will perform as directed, then submit a claim for directed acceleration, arguing that although it appeared from the original schedule that the builder would not complete on time, there was no risk of failing to complete on time at all. The builder might state that in the absence of the acceleration order it would have finished on time anyway.

The proper handling of a situation such as this is not to issue an acceleration order (even though acceleration is justified or needed), but to send a letter to the builder calling attention to the fact that completion by the scheduled date will be required, and that "according to the schedule it appears that the builder will be unable to complete the activity by the completion date indicated in the contract. A revised schedule shows how the builder plans to complete the activity by the scheduled date." This leaves the means and methods to the builder, as well as the responsibility.

Often the problem starts in the architect's design office. Project completion dates or time available to complete is frequently established by an architect or design professional who has had little field experience, and thus is hardly qualified to properly establish the real-world time duration needed to complete all the activities of the project. Typically one will see examples of this in activity durations that are too long or too short. If insufficient time is allotted to the completion of a project, all the bidders will be forced to bid the job as accelerated work, thus increasing the costs substantially. One indication that this may be happening comes when all the bids are in, and, although they are grouped close together (a good sign of competent documents), they are all considerably in excess of the architect's estimated project cost. This can be evidence that all of the bidders are bidding on the job as accelerated work. This will increase the cost of construction above that for the same job completed within a normal project schedule.

Another risk surfaces here as well. After figuring a bid for a job as being all accelerated work, a bidder may look in the contract documents to determine how much liquidated damages are being assessed in case of a completion delay. Upon finding daily liquidated damages to be assessed, the bidder refigures the bid to include that amount.

In the foregoing example, let us assume that we plan to build a project that on a normal residential construction project schedule should take 16 months to complete, but the contract documents actually indicate the completion must be within 14 months, including two months acceleration of work. Then the bidder will most certainly check the amount of liquidated damages called for in the contract documents. Let us presume the amount of liquidated damages was only $500 per day. An enterprising bidder will now go back to its 14 month bid and refigure the costs for completion in 16 months instead of the required 14 months. Then the bidder will determine that the difference in time that was allowed by the owner for completion versus the time that a job such as that should normally take was two months, or 44 working days. The bidder now multiplies 44 times $500 per day and adds the amount of liquidated damages to its total bid for doing the work in 16 months, and submits it as the bidder's bid price.

The problem here is that the cost of liquidated damages, at $500 per day, is considerably less than the daily cost of acceleration to the builder; thus, the bidder submits a bid claiming that it will complete on time, when in fact it planned from the very beginning to complete the work two months late and pay liquidated damages costs. The owner and architect turn out to be the only persons who were unaware that the project was destined to be late before it was even started. Generally speaking, any job that has a resident project manager operating out of an on-site field office can justify in excess of $1000 per day of liquidated damages without difficulty. Typically in commercial construction in California, this figure starts at $2000 per day and goes ballistically upwards.

A related condition called *deceleration* can also be experienced on a project's schedule. This occurs when a builder is directed in writing or constructively to slow down its job progress. Many of the same considerations that apply to acceleration also apply to deceleration. In preparing for a claim for acceleration or deceleration, it should be borne in mind that the costs to the builder for going into premium time, such as working an extended work week, cannot be computed simply as including the added hourly costs multiplied by the additional hours. Studies have shown that, as the work week is extended, there is an accompanying drop in worker productivity. Furthermore, as the extended overtime is continued, the productivity rate continues to drop. Scheduled overtime is not often seen on competitively bid lump-sum contracts, as most builders are well aware of the negative effects of overtime on cost and productivity.

Simple arithmetic shows that premium pay for double time or time-and-one-half makes overtime work much more expensive. However, people who insist on overtime seldom realize that other costs associated with overtime may be even more significant than premium pay. Premium pay affects only overtime hours, but continuing of scheduled overtime drastically affects costs of all hours. All available research indicates a serious inverse ratio between the amount and duration of scheduled construction overtime and the labor productivity achieved during both regular and overtime hours.

Studies have shown that, in the first few weeks of scheduled overtime, total productivity per person is normally greater than in a standard forty hour work week, but not as high as it should be considering the number of additional work hours. After seven to nine consecutive fifty- or sixty-hour work weeks, the total weekly productivity is likely to be no more than that attainable by the same work force in a standard forty-hour work week. Productivity will continue to diminish as the overtime schedule continues. After another eight weeks or so of scheduled overtime, the substandard productivity of later weeks can be expected to cancel out the costly gains made in the early weeks of the project's overtime schedule. This means that the total work accomplished during the entire period over which weekly overtime was worked will be no greater, or possibly even less, than if no overtime had been worked at all.

When the loss of productivity is combined with the higher wage cost (including premium pay), productive value per wage dollar paid, after several weeks of scheduled overtime, drops to less than seventy-five percent for five ten-hour days, less than sixty-two percent for six ten-hour days, and less than forty percent for seven twelve-hour days. When an overtime schedule is discontinued, it has been found that there was a dramatic jump in productivity per hour after return to a forty-hour work week. Construction delay claims involving acceleration of the work usually include claims for loss of productivity, which often exceed all other claimed amounts.

As an example, the following breakdown of a claim on a residential job illustrates the relative magnitude of the claim for loss of productivity, as compared with the other issues shown:

1. Extended project overhead	$ 10,019
2. Unabsorbed home office overhead	2,620
3. Labor escalation	1,430
4. Material escalation	1,329
5. Labor loss of productivity	8,424
6. Subcontractor claims	9,407
SUBTOTAL	$ 33,229
7. Profit and overhead on items 1 to 5	12,575
8. Unresolved changes	7,993
9. Interest on money	1,073
10. Additional bond premium	1,844
TOTAL CLAIM	$ 56,714

Contract Change Conflicts

This is an often misunderstood area in builder claims against the project. However, the probability of recovery by the builder as the direct result of such conflicts is good, insofar as the settlement is limited to the cost difference between the project cost as the plans and specifications are interpreted by the owner or architect and the builder.

Most construction contracts are called *contracts of adhesion*, which is a term applied to contract documents that are drawn by one party and offered to the other party on a take-it-or-leave-it basis, where there can be no discussion of terms nor contract modification by the other party. In the event of contract schedule conflicts here, however, the builder does have one advantage. In case of ambiguity, the court will interpret the contract in the builder's favor. This does not relieve the builder from the obligation of building the work in accordance with the interpretation of the architect or design professional, but only assures that the builder be paid.

Frequently, the builder will find that outdated standards are specified, or that products are specified that no longer exist. Often, the specification will contain references stating that, wherever codes or commercial standards are specified, the builder is obligated to use the latest issue of that standard existing at the time the project went to bid. Unfortunately, in many cases the designers considered the fact that the design was based upon an old standard that existed in their files, or a standard that was current during the design phase, but which later may have been updated by the sponsoring agency without the designer's knowledge.

Occasionally, serious difficulties arise from such practices, and the builder has the legal right to assess the project cost difference resulting from the error. Perfect specifications are hard to find, so the builder must make a reasonable interpretation at the time of bidding the job and will then be in a good position for recovery if a variance exists. The builder's interpretation must be based on what a reasonable person would interpret the documents to mean; then the builder will have the court on his side. A builder should never attempt to construct any questionable area, or change, without first submitting to the owner or architect a request for clarification, or notifying them of an error. This is known as an RFI (request for information), or an RFC (request for clarification). In most contract forms, failure to do this may bar full recovery of builder costs.

Other problems that often come up in production scheduling are directly related to the conduct of the owner's representatives on the job, such as the project manager or site supervisor issuing changes in the work that are of such magnitude as to constitute a cardinal change which creates a breach of contract. Or, on a lesser scale but indirectly related, is the case where one activity subcontractor may negligently delay another, which may result in the owner's seeking recovery from the negligent builder to pay the builder that was harmed.

Some of the types of scheduling problems that fall into this category include the following:

- Damage to work by other prime contractors in fast-track scheduling
- Breach of contract
- Cardinal changes
- Beneficial use of the entire project before completion
- Work beyond contract scope
- Partial utilization of the project before completion
- Owner's nondisclosure of site-related information
- Owner's failure to make payments when due

Some scheduling techniques, such as CPM or other network scheduling systems and their associated computer-generated printout sorts, have also compounded problems between the owner and the builder. Often the owner is swamped with stacks of computer printouts that he or she either does not want or does not know how to read. Yet somewhere in that stack of paper may well be the key defense against claims being made against the project and the owner. The question of scheduling methods of operation, and the impact of deviating from the anticipated schedule, requires a careful analysis to determine the reasonableness of the originally anticipated production schedule in conjunction with the planned methods of operation.

The analysis of every schedule-related problem is unique, and every such analysis should include a review of the anticipated sequence and schedule together with a review of the actual progress of the work. The review must include an analysis of any delays and impacts caused by all parties to the progress of the project.

It is in the presentation of the summary sorts, containing cost data and their supporting documentation, that most construction claims are the weakest. A carefully detailed reconstruction of all construction activities and their related costs must be made and presented so that they can be clearly identified for use in negotiating affirmatively or defensively. Small-volume construction companies are historically poor in their cost recordkeeping systems (or lack of systems), and they traditionally lack ability to assign costs to possible claims areas. An accounting firm is not the answer either, because often those cost items relating to potential claims are not identified by an accounting firm. Most accounting firms are not familiar with the cost-control procedures for construction contracts. A builder who realizes this and takes steps to structure and run cost-tracking and audit trail reports will provide his business with the best form of protection against claims. Remember, records are the builder's first line of defense in construction litigation documentation.

One of the most common causes of builder claims occurs during attempts to price change orders. Typically, owner change orders contain a waiver clause that requires the builder to guarantee that the price and time named in each individual change order represents the total cost to the owner for that change, and the builder waives any rights to impact costs. This, unfortunately, leaves the builder only one recourse—the claims process. The court has traditionally held that an owner may not force a builder to "forward price" its impact costs when pricing an owner-directed change. The court has held that while it might be good contract administration on the part of the owner to attempt to resolve all matters relating to a contract modification (change order) during the negotiation of the modification, use of a clause which imposes an obligation on the builder to submit a price breakdown required to cover all work involved in the modification cannot be used to deprive the builder from his right to file claims.

The production on a project can usually be suspended by the owner for any of several reasons. In each case, the owner or its agent must keep detailed cost isolation records of all activities affected by the suspension to defend against a claim. It should be kept in mind that suspension of the work for any amount of time, such that the completion date is extended, may impact the builders' costs through unabsorbed home office overhead and the real possibility of missing other projects due to the delay.

Builders may also claim the effect upon their organizations of the costs related to dismantling operations, mobilization or demobilization, direct costs, settlement expenses, escalation costs, prior commitments, post-termination continuing costs, unabsorbed overhead, unexpired leases, severance pay, implied agreements, restoration of work, utility cutoffs, inventory, replacement costs, and all other allocable costs. On suspensions of the work, be certain that all such orders are in writing and that a careful, detailed record is kept of its total effect on the builder's time and costs.

Chapter 6
Strategic Use of Change Orders

The Six Ps of Change Order Management

Despite their differences, all change order types will lend themselves to nearly the same analysis. The change order life cycle will progress through the same steps in virtually every case. The individual divisions in the process may be more or less clearly defined, depending upon the specific change. The relative importance of each part of the process will vary depending upon the category of the change. The objective for the contractor is to identify the steps in the process, and assess relative values to their importance. The next step is to proceed methodically through each one and then monitor performance in time to improve unacceptable results. These steps are traditionally called the six Ps of change order management:

- Prospecting (discovery)
- Preparing
- Pricing
- Presenting
- Performing
- Payment (getting paid for)

Change orders are an integral part of the residential construction business, especially the custom home market. Clients who are creating their dream home aren't going to think of every item and feature during the design stage. They will want to alter, delete, and add up until the final walk through. It's not unusual for the buyer of a $750,000 custom home to have $50,000 in changes. Given this situation, your ability to handle change orders effectively, efficiently, and profitably is vital to the success of your business.

The key to a successful change order management policy is to have one. It really doesn't matter what your policy is, as long as it works for you, and that you make the policy clear in your contracts from the start. If you lay out the ground rules in handling change orders in your execution of the contract documents and adhere to them yourself, most owners and architects will go along. Your change order management policy should have three basic factors:

- **Document everything**. Create a change order form that lists the details of the change, the date, the price, the time effect it has on the project schedule, the cumulative effect on the project's price, and the new projected closing date. Never make a change without a written form, and always have the owner sign it. Create a change order log and keep track of all change orders' status.
- **Price the change order carefully**. Never guess. Price change orders after you have gone through a step-by-step summary of its cost and its analysis of the impact on the total project schedule. Contractors often forget to include a wide variety of costs in their change orders, including sales tax, additional permit fees, special inspections, office overhead, equipment use, and personnel costs. Additionally, they aren't willing to take enough profit to compensate for the extra work that change orders require. Today's change orders are usually priced at your cost plus 100 %.
- **Schedule the delay carefully**. Builders can sometimes err even worse on figuring the delay caused by the change order. Determining the price for an item is fairly straightforward addition, but how do you know how much time it will take to change the linoleum in the laundry room? Put in a substantial time delay for any change order and if owners balk, don't give in. Tell them that is the delay and if they want the change they will have to sign for the delay.

Discovery

Identifying the potential for change orders is an important part of effective change order management. It boils down to breaking the project back into its component parts, looking back into the design and approval process to determine if it was properly or improperly completed, dissecting the contract itself to expose potential conflicts, and researching the contract documents to determine where the errors and omissions lie.

When searching out potential change orders, remember that changes of a given type usually do not occur very close to each other and secondarily, that the way one change order was handled successfully in the past may no longer apply because of differences in the people involved, contract provisions, timing, dollar value, or owner/architect dispositions. The potential for friction between the project's parties exists in every change order. This is because implicit in an identification of a potential change is an observation that some error or oversight exists, and someone did not do their job properly. If it is you who are preparing or are identifying the change, it will be wise to double-check all parameters. Don't fall into the traps laid by assumptions. Read and reread the contract documents to confirm or correct your original impression. Be sure (particularly the first time) that you're correct. This extra precaution will first establish and then maintain credibility in the introduction of future change orders.

Tactics

As soon as a contract change is recognized to have occurred, thoroughly research the contract documents to confirm the case. Pay particular attention to any notice requirements that may exist. Upon such confirmation, immediately prepare and submit a complete change order proposal. Take great care to ensure that it complies with all proposal requirements in every respect. Give the change order as much legitimacy as possible. Make the additional work seem obvious. Since the burden of proof of the owner's conduct is on the contractor, get the owner's conduct down in writing and included into the official project record. Remove at least that portion of risk.

We've seen that if a specification is subject to more than one reasonable interpretation, the contractor has the right to choose the interpretation. This is one of the strongest leverage points in change order dispute resolution, and should be known as such by every residential builder and contractor.

Change Order Preparation

Immediately upon the apparent discovery of a potential work change, it is important to record the event. Here it is important to understand that a change order file can always be voided if it is later determined that the item is contractual. Begin the documentation now. The potential extra paperwork is a small price to pay to eliminate the possibility of overlooking important facts in a fourth-quarter backward analysis. Worse would be letting a compensable change get shuffled back into the project to be lost track of entirely by the time the project is about to be closed out. An easy, systematic documentation procedure will prevent these kinds of costly oversights.

A separate general file is established to track all potential project changes through their eventual resolution. I use the word "potential" to reinforce what was previously stated, i.e., be sure that even questionable issues continue to receive attention until they are either confirmed to be contractual or until they get paid for. Within the general file are individual files for each potential change. These individual files will contain all important documentation, including date of discovery; all circumstances of discovery; person first responsible for discovering the change; all written documentation (reports, invoices, letters, meeting minutes, schedules, memos, note, etc.); plans, sketches, specifications, diagrams, etc.); and photographs.

The general file should also contain a separate section for a change order log. This is a summary record to maintain a clear overall visibility of the current status of all changes in the work. A good change order log will allow for fast analysis of apparent trends in the project and important evaluation of the timing and success of the change orders' administration.

Legal Aspects of Time in Change Orders

The court historically has not used consideration of time in project schedules as affecting judgments in construction litigation lawsuits. This is due to tendency in recent decades for frequently delayed construction projects to end up in litigation. Timely completion depends upon proper performance by the many participants, as well as optimal conditions for performance, such as weather. Nowadays, encountering unanticipated subsurface conditions is the kiss of death. Inspections, excavations, and soil remediations are just a few of the costly delays that can result.

Delayed performance, due to encountering unanticipated subsurface conditions, has been held by the courts to be invalid grounds for termination of the contract. This means the owner and project team continue to eat it financially. Delayed performance is also unlikely to create legal justification for the owner's refusal to pay the contractual compensation. At this point, the owner is typically left with inadequate damage recovery. Insurance policies and bonds to protect damage recovery in such cases are extremely expensive and typically put the project out-of-budget and therefore make it not viable.

Similarly, delayed payment by the owner is less likely to automatically give the contractor a right to stop the work or terminate its performance. Finally, a strong possibility exists that a performance bond that does not expressly speak of delay will not cover delay damages. These damages will then be contested, and it falls to the contractor to prove the validity and logic of the project schedule.

Computation of delay damages is also different. If the owner does not pay or the contractor does not build properly, measuring the value of the claim is, relatively speaking, simple. If the owner does not pay, at the very least, contractors are entitled to interest. If the contractor does not build properly, the owner is entitled to the costs of correction or the diminished value of the project. Delay creates serious measurement problems in project schedules. The owner's basic measure of recovery for unexcused contractor delay is lost use of the project. The contractor's basic measure of recovery for owner-caused delay is added expense.

Lost use is difficult to establish in noncommercial projects. Added expense is even more difficult to measure. Because of measurement problems, each contracting party, whether it pictures itself the potential claimant or the party against whom the claim will be made, would like a contractual method to deal with delay claims, either to limit them or to agree in advance on amounts. This does not mean that time is not important in construction. The desire to speed up completion, due to money investment pressures, is what motivates the project owner.

Traditionally, two schedules are initially purposed for any project. The general contractor supplies a version suited to his company's timing for profit, and the owner supplies one suited to his timing for profit. The key factors for these two schedules to merge successfully is how close each is to the critical design, procurement, and construction of all aspects of the project. Once the schedule is completed, it needs to be cooked. This term means a critical analysis of the schedule by the contractor and the owner. At this point, the schedule is debugged by using "what-if" scenarios through the critical paths of production. These first editions of the schedule are generalized overlays, called macros. Macros are then redesigned to reflect sequential activities' interdependencies and events time scaling. Important events and individual phase completions are noted as milestones. Owners typically tend to be overly optimistic about all activities happening with no problems, their schedules showing an unrealistic and often unachievable project completion date. On the owner's schedule, activities are strung together in a tight series, with each activity closing event also being listed as the starting event for the next activity. In this type of configuration, all activities lie on one critical path. This makes for a network diagram with little or no flexibility in scheduling float allowances, which is the primary area a contractor makes adjustments in. General contractor's schedules, conversely, show free float time in their subcontractors' related activities to serve their own work schedule, usually with no free float linkage to overall project total float. Obviously, with what you have already learned about the vital management uses of float, you can see that neither of these two schedules will be adequate. What invariably happens in a real-world field application is that the initial activities burn up all available total float, and any emergency float time is then borrowed from the succeeding activities, leaving little or no float time along the critical path. This always results in the critical condition of production and cost overruns, because contractors do not want to lose the bid and will often accept an unrealistic deadline, hoping that they can pull a rabbit out of the hat somewhere along the line.

AIA Document A201, Section 3.10.1 requires the contractor to submit its construction schedule for the information of the owner and the architect. The schedule must provide "for expeditious and practicable execution of the Work." The contractor's failure to conform to the most recent schedule constitutes a breach of contract under Section 3.10.3. Contracts prepared by experienced public or private owners, particularly private owners under the influence of their lenders, usually prescribe much greater detail and take greater control over the contractor's schedule.

This can manifest itself in language requiring that the project schedule be on a form approved by the owner or the owner's lender; that each monthly schedule specify whether the project is on schedule (and if not, the reasons therefore); that monthly schedule reports include a complete list of suppliers and fabricators, the items that they will furnish, the time required for fabrication, and scheduled delivery dates for all suppliers; and that the contractor hold weekly progress meetings and report in detail as to schedule compliance. Similarly, the Engineers Joint Contract Documents Committee (EJCDC) takes progress much more seriously than does the AIA. For example, the EJCDC's number 1910-8, Section 2.6, requires the contractor to submit, within ten days after the effective date of the contract, an estimated progress schedule, a preliminary schedule of values, and a preliminary schedule of submittals. The finalized schedule must be acceptable to the engineer, and Section 6.6 requires the contractor to submit, for acceptance, adjusted progress schedules. The approved finalized schedule becomes a contract document.

Outside Delays

Most field daily reports have columns for weather delays. An inclement weather system passing through your area that turns out to be a toad floater will also delay outside activities, such as excavation, grading, paving, and placement of concrete. These are considered constraints of practical limitations and must be factored as such in the project schedule. The reasons are twofold. First, time cushions must be installed within total float, to provide allowances in practical production parameters. And second, AIA Doc A201, Section 4.3.8.2 (which is the section dealing with claims) requires the contractor to document "by data substantiating that weather conditions were abnormal for the period of time and could not be reasonably anticipated" before the contractor can receive a time extension for weather conditions. Section 4.3.8.2 further requires that the contractor document that the weather conditions "had an adverse effect on the scheduled construction."

The overall precedence concerning schedule outside delay is found in AIA Doc A201, Section 8.3.1, which states: "If the Contractor is delayed at any time in the progress of the Work by any act or neglect of the Owner or the Architect, of an employee of either, or of a separate contractor employed by the Owner, or by changes ordered in the work, or by labor disputes, fire, unusual delay in deliveries, unavoidable casualties, or any causes beyond the Contractor's control, or by delay authorized by the Owner pending arbitration, or by other causes which the Architect determines may justify delay, then the Contract Time shall be extended by Change Order for such reasonable time as the Architect may determine."

This section delineates the parameters of acceptable and enforceable delay claims submitted by the contractor. These requirements can make it difficult for a contractor who does not keep detailed weather records to claim a time extension for adverse weather conditions. They reflect a belief by the AIA that weather generally is a risk assumed by the contractor, and that only in extraordinary circumstances should weather be the basis for a time extension. The legal documents used by the courts to determine judgments in weather related claims are a Daily Inspection Report (DIR) and a Quality Assurance Report (QAR). If the owner is unaware of these documents, then certainly you, the professional contractor, should be. And you should implement them on your behalf to protect your schedule.

Another reason for not granting a time extension is the single contract system's objective of centralizing administration and responsibility in the prime contractor. Only if subcontractor-caused delay is specifically included will it excuse the prime contractor. This needs to be stated within the contract specifications to protect you from the schedule's default through circumstances beyond your control. The independent contractor rule, though subject to many exceptions, relieves the employer of an independent contractor for the losses wrongfully caused by the latter.

As noted previously, prime contractors assert that subcontractors are independent contractors because the subcontractor is usually an independent business entity and can control the details of how that activity is performed. Even so, the independent contractor rule does not relieve the employer of an independent contractor when the independent contractor has been hired to perform a contract obligation, and, under that employer's direction, the party suffering the loss caused by the independent contractor is the party to whom the contract obligation was owed.

Legal Notices

During the project's scheduled time line, certain legal notices must be filed and tracked by the prime contractor. The primary reason for this is litigation protection from claims of nonpayment of services.

When an activity subcontractor is not paid by your project's prime contractor, that subcontractor can file a mechanic's lien and/or stop notice at any time after ceasing to perform labor or furnishing materials, and until thirty days following the notice of completion or issuance of a notice of cessation.

A prime contractor, on the other hand, if not paid, can file a mechanic's lien within sixty days after notice of completion or issuance of a notice of cessation.

A schedule of these legal notices filing dates should be included in the builder's finalized schedule. Let's look at each legal notice closely.

Notice to owner

The contract document, *notice to owner*, describes in non-technical language, in clear and easy to understand words, the state's mechanic's lien laws and the rights and responsibilities of both the owner and the contractor. This document also advises the owner to require payment and performance bonds from the contractor to protect the owner from the contractor's not finishing the job and not paying his subs and suppliers. A bond is an insurance policy that has been taken out to guarantee a certain event will happen or not happen.

A performance bond is an insurance policy underwritten by an insurance company to ensure that the contractor will perform the work as contracted. If the contractor abandons the job, the insurance bond will cover the cost of another contractor's coming in and finishing the job. A payment bond likewise guarantees the owner that the contractor will pay his subs and suppliers. This document must be given by the contractor to the owner before he or she signs the contract. If it is not included or attached to the original contract, the contract is invalid. This is a legal technicality big enough to nullify claims made against the owner or project.

Stop notice

A stop notice is filed by suppliers or subcontractors who have previously filed preliminary notice in a timely manner, to the holder of funds (public agency, owner, or lender). It is a legal notice to request that funds be withheld from the general contractor. Stop notices can be sent on public or private jobs. The stop notice is used on public works jobs in place of a lien, because public property cannot be liened. Any activity contractor, except the original (prime) contractor, may serve a stop notice.

A stop notice is a "lien on funds." It is a legal notice to the awarding authority or the holder of funds by a subcontractor (claimant), to withhold from the prime contractor in order to satisfy claims for labor and materials furnished to the prime contractor. When a subcontractor (specialty contractor) has contact directly with anyone from the awarding authority (government agency), the subcontractor is then the original (prime) contractor and is subject to the provisions of the law relating to original contractors.

Stop notices must be verified by the owner to be true and correct. However, the claimant on a stop notice is the contractor, not the owner. To obligate the holder of funds to withhold funds, the stop notice should be bonded for 125% of the amount of the claim. This is called a bonded stop notice and it is the heavyweight champion of contract documents. A stop notice stops the flow of money into a project, and things get critical quickly. Upon receipt of a stop notice, the lender may withhold funds from the owner or prime contractor.

Upon receipt of a bonded stop notice, the lender must withhold funds from the owner or prime contractor.

Both require a lawsuit to force actual payment. Neither mechanic's liens, notices, nor payment bonds impair the contractor's right to court, but rather are legal aids to recovery. The stop notice is used only by subcontractors and third parties, such as suppliers. It is usually more effective than a mechanic's lien because it is used to withhold a specific amount of money from the prime contractor, which the subcontractors claims he or she is owed. Since the stop notice is used to force the lender (or holder of funds) to withhold a disputed amount of money, the subcontractor can be assured that, if any money is left in the account, it will be held until the dispute is resolved.

A stop notice is the contractor's only single means to force payment on a public works job. Public property cannot be liened, so in the event that a contractor needs to serve a stop notice on a public works job, he must send it to the awarding authority which is the project coordinator for that public works job. A bonded stop notice ensures compliance by the construction lender to withhold funds from the prime contractor to satisfy the amount claimed by the subcontractor or supplier. Without the bond, the holder of funds is not obligated to withhold the money.

Notice of completion

Notice of completion is normally filed by the project owner within ten days after the project is completed. This legal notice restricts the number of days within which a contractor can file a lien against the property. This should also equate to a milestone in your schedule that marks the ending event of the project production and the starting event on project closeout.

Once this is filed, subcontractors should do no more work. The courts have held the criteria to determine the completion of work to be:

1. The occupation or use of a work of improvement by the owner.
2. The acceptance of work by the owner.
3. If there has been a cessation (stoppage) of labor on the job for a period of 60 days without a notice of any kind being recorded which constitutes a legal completion.

Notice of cessation

This is the legal notice used if, for any reason, the job has been stopped because of a union strike, trucking strike, labor dispute, labor shortage, or a similar reason. It cannot be recorded until there has been a cessation of labor for thirty days, with the exception of Force Majeure.

Notice of non-responsibility

This notice is filed by a project owner in a conspicuous place upon the property within ten days after obtaining knowledge that work that was not ordered is being done. This notice must also be recorded with the county recorder in the area of jurisdiction wherein the property is located, giving notice that the owner will not be liable for any accident or injury occurring.

Preliminary notice

Preliminary notice is a contract document filed by subcontractors and suppliers within twenty days of beginning their activity on the project. On residential and commercial jobs, preliminary notice is sent to the lender, the owner, and the prime contractor by registered or certified mail, or is delivered in person. On public jobs, legal notices are sent to the awarding authority representing the public agency and the prime contractor. Preliminary notices are filed to establish the subcontractors' and suppliers' rights of lien and stop notice. Subcontractors and suppliers on public works jobs do have stop notice rights in the event of delay in progress payments. However, if preliminary notice is filed late on a public works job, all stop notice rights of the contractor are lost.

Preliminary notice must be filed by subcontractors and suppliers within twenty days of their beginning work on their specific activity within the project schedule. Therefore, it is mandatory that every licensed contractor doing any activity within your project schedule, with the exception of the prime contractor, give its notice not later than twenty days after it has first furnished labor, services, equipment, or materials to the project's jobsite. The prime contractor does not give a preliminary notice, due to the fact that its contract with the project owner is the first, or prime, contract and serves as preliminary notice.

The purpose of preliminary notice is to notify those vested parties of the project who never have a direct relationship with (and often never see) the activities subcontractors who do the work. One of your reports should track submission of these legal notices, and one of your time line sorts should track their timetables. Contract law statutes require that if any work is done by any of your subcontractors for over $400, the subcontractor is subject to disciplinary action by the Contractors License Board for not filing this legal notice. If notice is given later than twenty days on a private job, the contractor may not recover the value of work done at more than twenty days of his activity construction from the date of the notice. Lien rights are lost as to anything furnished more than twenty days before serving of the notice.

An activity subcontractor not having a direct contact with the prime contractor, other than a laborer, must, as a necessary first step to the validity of its stop notice rights, give a written notice to the prime contractor and the public body (awarding authority) concerned, within twenty days after it has first furnished labor, services, equipment, or materials to the jobsite. This notice takes the form of preliminary notice.

Contrary to the law on private jobs, if the claimant on a public works job lets his first twenty days of activity work elapse without filing preliminary notice, said contractor loses all stop notice rights forever. But that contractor still has a legal right of action and can sue against the bond and the bonding company, since all public works jobs are bonded. These public works bonds protect all those involved in the project construction from non-payment for their services. A 90-day public works preliminary notice is necessary to enforce a claim on a payment bond on public works jobs. It is required in public works projects by subcontractors and suppliers to the prime contractor, within ninety days after furnishing labor, materials, or services.

Mechanic's lien

A mechanic's lien is placed upon the project's real estate to secure the compensation of those who have been directly instrumental in the improvement of such property. This is a critical factor of final payment for the services of a professional contractor in that, should the project go over budget and subcontractors or materials suppliers not be paid in a progress payment, mechanic's liens filed may halt the production process. The project may wind up in litigation and your services may go unpaid. In such a case, you may need to file a mechanic's lien to secure payment for your services.

Wage earners do not have lien rights, but materials suppliers, contractors, and laborers have a lien upon the property upon which they have bestowed labor or furnished material for the value of such labor done and materials furnished. The legislature provides, by law, for the speedy and efficient enforcement of such liens. Mechanic's lien law is of particular interest to all persons in the building and construction industry, inasmuch as the value of labor performed and materials furnished to any building or construction project are debts, the payment of which can be secured by a lien binding the real property as security for the value of the labor performed and materials furnished. A suit to perfect claim of lien, or *lis pendens* in lawyerese, binds the real property for the value of the labor performed and materials furnished similar to the manner in which a mortgage or deed of trust secures the payment of a loan or note. A claim of lien means a written statement, signed and verified by a claimant or his agent, containing all of the following:

- A statement of his demand after deducting all credits
- The name of the project's owner or reputed owner, if known
- A general statement of the kind of labor, services, equipment, or materials furnished
- The name of the person, owner, or developer by whom he was employed
- A description of the jobsite sufficient for identification

A claimant is one who has made any site improvement at the request of an owner or his agent (architect, engineer, prime contractor, etc.) and has a lien upon such owner's property for work done or materials furnished. The first step in establishing a lien right is for anyone (with the exception of the prime contractor or workers working for wage) interested in establishing a lien right to file a preliminary notice. If an invoice from a material supplier contains the required information, it may serve as preliminary notice. Mechanic's liens are rights under each state's constitution and are liens on the property for "value of the labor done and materials finished." A mechanic's lien is a legal means of recovering payment for works of improvement and site improvements by contractors, subcontractors, laborers, and suppliers of materials, including, but not limited to, new construction, construction alterations, tenant improvements, additions (works of improvement), repairs on almost any structure, or demolishing or removing improvements (site improvements). Mechanic's liens attach to both the structure and the land on which it sits. The requirements for mechanic's liens against the project are:

- Commencement of work, however slight. There must be actual visible work, and/or delivery of materials to the jobsite.
- It must be shown that labor and materials became a fixture on the land, and that they are of a structural and permanent nature.
- Work must be done at the request of the owner, or his agent, contractor, subcontractor, architect, builder, etc. Knowledge by the owner of the construction is grounds enough for judgment.
- If an owner of a structure or property records a notice of non-responsibility within ten days after knowledge of the construction, no mechanic's liens will apply against him unless he is the one ordering said work to be done.

Liens are filed with the county recorder's office in the county and jurisdiction in which the project is located. Liens can be filed by all original contractors, subcontractors, or suppliers who have previously filed preliminary notice. Preliminary notice is a required prerequisite by law in order to file a mechanic's lien. These liens must also be verified by the claimant (the contractor or supplier) to be true and accurate and if not correct, are considered void, therefore unenforceable by the court. They must be filed no later than thirty days after notice of completion by subcontractors and no later than sixty days for prime contractors. If no notice of completion is filed, then all have ninety days from their finish date to file. Once filed, liens must be perfected in ninety days from the date the lien was filed. This is called a "suit to perfect lien" and it forces foreclosure of the property by court judgment of default of payment. However, the claimant is allowed by law to grant extensions of the foreclosure for up to one year.

Just having the mechanic's lien right and filing the required Preliminary Notice in a timely manner does not guarantee that the contractor will get paid. It simply means the contractor has what is termed legally a "cloud on the title," the same as a mortgage or grant deed holder. For the contractor to get its money, it must do exactly what a bank would do: Foreclose. Foreclosure requires a lawsuit to perfect a particular Claim of Lien, and it must be commenced immediately after the date of filing of the particular Claim of Lien. Liens and the perfection of liens must be filed with the county recorder in the area of jurisdiction where the project is located. If the contractor is late on this, the contractor loses. In fact, if the contractor is late on any of these legal notice filings, the contractor loses. The courts and construction law hold a licensed contractor to a higher level of accountability and responsibility. A prime contractor can't stand in court and say, "I don't know, I must have overlooked my filing dates."

For a licensed contractor to get its money and perfect a lien, it is necessary to file a lien foreclosure action within ninety days of recording the claim of lien. A suit to perfect lien, which establishes a court date to sue for foreclosure on the property to settle claims of unpaid money due the contractor, is this necessary second step. If more than ninety days are needed in which to foreclose, credit should be given to the project owner. Within ninety days after his offer of extension, a contractor must foreclose or file for an extension. The contractor is the only entity that has the right to give the owner an extension for up to one year. Recorded extensions may be granted to the owner by the above procedure, but the contractor must foreclose within a year after work is completed. The action should be brought to trial within two years after commencement of suit to perfect lien.

Cost Tracking

One of the many crucial tasks of contract change order management for contractors during project operations is that of monitoring the project field service costs, as separate and distinct from the architect's design costs. The amount of the field costs is a function of the size and classifications of the on-site field labor forces assigned to the project, field office overhead costs, materials and supplies, support services from the home office, vehicle leases and fuel charges, the field office's share of corporate general and administrative (G&A) costs, outside consultants or contract services, and job profit.

By tabulating the monthly accumulations of budget and actual field service costs and then plotting each amount on a time-versus-cost chart, similar in form to the traditional S-curve chart, you can make a visual comparison that clearly shows not only the status of the contract at any given time, but will also show the project manager or owner any change in trend toward either a savings or a cost overrun. By also plotting a curve representing the amounts invoiced to the owner for such field services, you can provide an additional dimension of usable data to the owner.

To prepare and maintain the cost-tracking chart on the computer program, enter the items into the cost by activity number program file. This sort will calculate the estimated, current, and projected due costs for every item you enter, in every activity you're using. Accurate data for your cost tracking requires regular inputs from the site supervisor on all field costs, and hours of work in each classification at the project site. The prime contractor arranges for all such field data to be received from the project manager at the end of each week. Typically, the monthly pay estimates of the project's subcontractors' work for progress payments are submitted by the 25th of the month, and all submittals are to be in the office for payment before the end of the month. Billings from the architect to the owner, however, usually are based on the closing date at the end of the month. Therefore, weekly tabulation of these data should not interfere with the owner's review of the subcontractors' pay requests. The cost tracking should be done on a unit basis in all activity items. These are then used to produce the project's audit trail.

One of the major advantages of computerized scheduling is the ability to link items' and activities' cost to the network schedule milestones of payment. Most major retail computer programs link cost tracking with inventory control (the barcode on the products at the grocery read by laser scan and entered into the computer for readout). For example, as the item is sold to you in the grocery store, it is also subtracted in the inventory. Milestones exist within such programs to signal the purchasing program to buy more of those items when inventory levels are low. All such information provides an audit trail of each item.

Audit Trail

The factors used by the computer to build a cost-tracking database used in computations of the project's audit trail are the activity's assigned *i-j* number; the activity description; start event of the activity; finish event of the activity; line float of the activity; available (committable) float; negative float (if any); prior event; next event; and the activity cost expense data (estimated, current, projected due).

The computer uses the float data entry to provide calculations under the "what-if" pulldown menu. This forecasting advantage allows the project contractor to input other numbers into the cost and time columns for computation. The next step is to critically analyze what those changes would bring if initiated immediately. The subtotals of each column can either be totaled or averaged, or both. Printouts can be run at any time, to freeze a particular cost scenario that appeals to the owner.

After critical analysis of all options is complete, and the prospecting for potential change orders is done, the program returns to the original numbers without disturbing any other data. This is especially useful in construction-loan interest-payment scenarios. These financial wraparounds allow the owner to maximize the potential of the loan funds without dipping into personal lines of credit, thus keeping working capital on hand and available. This overall process is known as *trends analysis.* By entering the relevant data on the cost by activity number sort, the computer now links the activities network with the costs and expenses necessary to run the network.

In addition, owners using CPM systems have the advantage of control over those costs before they are paid, by having the project contractor run trends analysis forecasting. The critical art of monitoring the project field service costs now linked to the project schedule allows you to separate those cost expenses from the architect's design costs. The cost by activity number sort tracks and calculates the classifications and the amounts of the on-site field labor forces assigned to the project, as well as the field office overhead costs. An entire database can be built to show costs of materials and supplies, support services from the home office, vehicle leases and fuel charges, field office share of corporate general and administrative (G&A) costs, outside consultants, professional legal services, professional engineering or contract services, and separate or combined cost-to-profit ratios.

Audit trail trends analysis is an important function of schedule estimating that will provide you with one of the prime advantages of computerized project scheduling, that of cost control and the ability to foresee change orders, and to move the commitment of resources before they are needed.

Chapter 7
Generating Change Orders

Prospecting for Change Orders

Locating potential change orders is a matter of locating gaps and omissions in the production schedule or project operations. This is done by an end-around perspective. First do a complete breakdown using the following standard estimating procedures, then look for any missing items, tasks, or activities in the contract general conditions or specifications. Look for a "missing link in the chain," if you will. This is called *costing strategy*, an analysis of examining the required expenses and locating overlooked or missing expenses. Construction estimating is an art, not a science.

Even the best of architects are only human and make mistakes. Their estimates for the owner have been known to be off on many occasions. Ask any builders you know if they've ever had to build using prints that were revised during construction. I can tell you they're going to say yes. On many residential jobs the range between high and low cost estimates supplied by the designer to the owner will be 20% or more. Once let out to bid, there's room for legitimate disagreement on what the correct costs are, even when complete plans and specifications are available, the date and site are established, and labor and materials costs are the same for all bidders.

No cost analysis fits all jobs. Finding potential change orders requires that an accurate estimate by the contractor is done first. Good estimates are custom made for a particular project and a single contractor through business judgment, careful analysis, and experience. Remember, claims consciousness begins with the knowledge that even seemingly insignificant problems can escalate through neglect or mismanagement to sizable liquidated damages. The final step after analysis of the estimate versus the specifications is to apply claims consciousness in preparation or prevention of situations leading to change orders and construction claims. This is done by careful analysis of errors and omissions, schedule delay, schedule acceleration, interference, and latent conditions. Each of the following costs need to be examined using these factors.

Generating Change Orders

"Improved information" change orders will normally be initiated by information that was not available at the time of bid document preparation. Previously unknown conditions may become evident or newer methods to achieve a function may be developed. These types of changes will result in a contract scope change that will be an easily approvable change order type. The caution here is that the guise of "improved information" may be used as a justification to cover up what is really an incomplete design or defective specification. This in itself is not necessarily bad. It may actually work in the contractor's favor as a face-saving mechanism. It may make it easier for the designer to justify the change with the owner with a minimum of criticism. A shrewd contractor will spot such an opportunity to remove the friction out of the change order acceptance and will get paid for the extra work while keeping the peace.

Despite their differences, all change order types will lend themselves to nearly the same analysis. The change order life cycle will progress through the same steps in virtually every case. The individual divisions in the process may be more or less clearly defined, depending upon the specific change. The relative importance of each part of the process will vary depending upon the category of the change. The objective for the contractor is to identify the steps in the process, and assess relative values to their importance. The next step is to proceed methodically through each one and then monitor performance in time to improve unacceptable results.

The first step in the generation of change orders is the clear identification of an item of work that is beyond the scope of the original contract. The origins and the timing of their discovery will ultimately determine the net impact on the contract work, as well as the ease with which it will be paid. It's easy to discover additional work when you're suddenly faced with it. Interruption is imminent (or has already occurred). Immediate resolution becomes critical if the most dramatic impact on both the contractor and the owner is to be avoided. Last-minute efforts then create a tension that minimizes options and clouds thinking. Interestingly, even though this scenario doesn't seem to make much sense when we sit back and coolly analyze the process, it is precisely this pattern that is allowed to happen over again by most contractors.

The real talent in prospecting for change orders lies in the ability to expose extra cost items in good time to allow their incorporation in stride. If we can learn to take the initiative to actively search and discover potential extra cost items, there will be time enough to work the process through in a less hurried manner. We'll have the luxury of being able to contemplate the best strategies and to consider those tactics that will get our change orders approved with a minimum of friction. It's the old business of making it happen instead of letting it happen.

On the other hand, there may be instances in which a last-minute presentation to the owner will be calculated to force a quick decision. If this kind of action is appropriate, it will only be when *you* have had enough time to plan. As such, you make it a quick (presumably less thoughtful) response on the part of your opponent. This approach is often employed in the legislature, for example, where literally hundreds of bills are introduced in the last days of a session. Quick action becomes forced without the opportunity for extensive research or debate. Tactically, it may be an advantageous approach if *you* have had the time to prepare, and not your opponent. Changes that are owner- and architect-originated are easily identifiable, but easy payment for these changes is relatively rare. The first step in securing payment for changes is always up to the contractor. The contract documents must be systematically researched to discover where the changes are.

Developing Change Order Precedence

The first step in developing the precedence of likely change orders is to establish the "levels of specificity." It is important that you decide at the outset where you will put your planning energy. Consider which parts of the schedule conflict and which aspects of the project are most important to your completion objectives. Set some general time expectations for each activity of the schedule and try to identify the job logic. This is a critical stage of the planning process, as you must answer each level of specificity question to avoid getting bogged down in the process or devoting an unreasonable amount of time to examining less important parts of the schedule. In the project schedule planning process, consider all of the following in creating a level of specificity in developing your change order precedence:

- Sufficiency of past performance data. How good is your company's historical information? Have the numbers shown a consistent pattern? If so, you can rely on them more; if not, find a better predictive source or database.

- Study vulnerabilities. Are there predictable risks? Do certain types of activities show poor event estimates in the past? How about the subcontractors? Are other construction projects similar to yours experiencing problems in areas of production?
- Seek opportunities. Are there particular areas of the project, such as atypical activities, which would benefit from special attention or critical inspection before scheduling? Find the greatest opportunity areas and devote extra time to them. These are areas in which you can save the owner money and thereby improve your professional stature and reputation.
- How much can you gain? What is the potential for increasing production in each activity? If the potential is small do not spend much time on it, even though solving the problem might give you some personal satisfaction.
- Time management. How much time can you devote to planning and developing the schedule? What resources can be economically devoted to developing the schedule? This is often a tough balance. The answer here lies in what kind of business person you are. If you spend much of your business time in fighting business fires between subcontractors' work schedules in developing the schedule, planning is most important but least convenient. Consider your personal business tendencies: are you generally obsessed with detail? Better then, in the interests of time management, to be a little more general to ensure getting the job done.
- Constraint. A constraint is a potential real-world limitation, or a tactic creating a float window during monthly recycling of the schedule, that can delay the starts of activities or tasks. Constraints, known as a dummy activity on network schedules, are shown as dashed lines with zero elapsed time. Constraints are negative factors that balance job logic calculations.
- Job logic. This is sequential relationships between activities, identified and defined during preschedule planning. These relationships consist of the necessary time and sequential order of operations throughout the project.
- Logic diagram. Arrow diagram of complete project network schedule, or a cross section of an area of production.
- Time-scaled chart. Logic diagram with a time scale.
- Velocity diagram. Velocity diagram means a straight-line, time-scheduled flowchart. The purpose of the velocity diagram is to determine the most efficient paths to joining the activities in total operational network scope.

Traditionally, the arrows in a velocity network diagram represent the activity itself, not the direction of movement. The angle of slope and arrow's length are not factors in the scheduling, simply the designer's choice. Each arrow in a velocity network diagram represents an activity, identified by its activity number.

- Velocity network schedule. This is the culmination stage of scheduling. All the preceding scheduling elements are computed into a master network plan for the entire project's scheduling, including post-project close out. Here, each activity is now also assigned its relative *i-j* number.
- Event diagram. The most common event system uses circles at the ends of each activity arrow. These circles are placed at the junction of the arrows and they represent the event, or the moment of time at which an activity is just starting or finishing. Important events which mark the completion of a phase, such as a foundation final, are called milestone events for they serve as benchmarks to the schedule's progress versus run time.
- Early start date. Earliest date an activity can start.
- Late start date. Latest date an activity can start.
- Early finish date. Earliest date an activity will be completed without float.
- Late finish date. Latest date an activity will be completed without negative float.
- *I* - *J* number. The letter *i* designated symbol for the tail of an arrow (start of an activity), and the letter *j* designated symbol of the head of an arrow (finish of the activity).
- Float. Measure of available scheduling leeway time on any activity's completion.
- Free float. Time by which activity finish event can be delayed without affecting the succeeding activity's start event.
- Total float. Measure of available spare time or scheduling leeway available on all activities' completion. Total sum of all free float.
- Negative float. Time a critical activity is late meeting its finish event.
- Phases. Groups of activities that will happen in a logical order, precede or succeed one another or happen simultaneously. As tasks are the micro units, phases are the project's macro units. Phases are arranged in divisions with sequential velocity management. Phases are supposed to flow into each other smoothly, on time, with no problems. If problems are anticipated, contingency planning in the form of workarounds need to be established ahead of time.

For example, in the time-compressed business world of today, these four basic planning steps often are completed simultaneously. Those trained in computer programs that "wash through" this data instantly to each stage, and change datum in all related stages if one factor is changed, using "what if?" scenarios. For the sake of instruction, though, we will use the assumption that they are each treated separately and in the order listed above.

Time-line computations are just simple addition and subtraction. The task of time-scaled computation is usually just as simple and easy. However, managing the timelines of the many tasks and activities involved in a modem project quickly becomes an enormous challenge. You will see that there are days when your schedule time line feels like a nervous mustang ready to leap out of control. Accordingly, modern competitive contractors use computers to manage their time lines. And with the use of dedicated software programs, the time line can be updated regularly without redoing the entire network diagram. Complex time variables can be "best line linked." Forecasting by use of trends analysis shows probability of various outcomes. Each activity can be updated regularly for accurate data factoring.

Typically, this is done in daily updates. This frequency serves two purposes. First, it guarantees timely accuracy in your data and pries open the largest window of opportunity necessary to make changes and workarounds before problems start costing money. Second, it produces a superior kind of detail orientation, combined with fast results, that tend to prevent budget cost overruns and give the astute contractor a definite advantage over the competition. This is the type of professionalism that produces profit and builds the reputation.

The complex scheduling problems that occur in multiresidential projects can be worked out rapidly by using program logic in a dedicated software network system, and time management becomes a viable tool for working out options, workarounds, and solutions. A programming task that normally takes hours to complete, if done by hand, can be accomplished in seconds by computer programs incorporating program logic. This rapid access to data is one of the key reasons for the successes of computerized scheduling. This speed of program logic allows variables to be worked out or planned around before they cost money. The enclosed accompanying software is configured using velocity network diagramming with program logic.

Scheduling Contingencies

Scheduling controversies often arise in the construction process, not only from disagreement over poor planning in the primary stages but also from operations among construction methods or sequences of activities and the project team and activity subcontractors in the secondary production stages.

Often these disputes occur where a project is constructed under multiple prime contracts, such as those commonly experienced in fast-tracking, phased construction, or in a professional construction management contract. The absence of any contractual relationship between the individual subcontractors and the architect has been the main cause of numerous legal disputes.

A cursory routine approval of contractor-prepared schedules can be the most costly mistake an owner can make. In the absence of a thorough review by the prime contractor, contractually required schedules can become the primary source of documentation from successful contractor claims. The contractor can simply provide as evidence in court its prebid, preconstruction, and progress schedules, and then compare them with the as-built or adjusted schedules.

The comparison can be very graphic, especially to those who have not had the training to know how computerized CPM works and how it can be subverted. If the owner approved the contractor's schedule, the owner can be left in an extremely weak legal position. This kind of comparison, to laypeople, is simple and straightforward. It will have a significant impact on a panel of arbitrators or a jury because it appears to be the truth, pure and simple. (Although the truth is rarely pure and never simple.)

But as we've already learned, the truth is rarely pure and never simple. Computerized network schedules reflect those who produce them. Prime contractors who understand the nomenclature and advantageous uses of computerized project schedules also realize how it can be abused. The only thing between the contractor and bankruptcy in scenarios like this are adequate records and a bulletproof project schedule.

So the prime contractor must know what scheduling contingencies are, and how to plan for them in the best interests of the client. Claims against the owner in a computerized CPM network schedule are, far and away, most often generated from changes in the schedule during production operation phases.

Changes in Schedule

Any scheduling change can have an impact on total project operations, and can even push back the contractual completion date. When scheduling changes are implemented as a result of constraints created by the owner or architect, the basis for a contractor claim may exist. Whenever the owner or architect issues any change order, field order, or work directive, or whenever a constructive change condition exists in any of the following areas, the contractor will begin to consider the impact of such action by the owner or architect upon its project costs and profitability.

These are the typical schedule contingencies that need advance alternative planning. They can lead to schedule changes that can later develop into owner- or architect-caused delays:

- Requirements to deviate from the schedule
- Orders to expedite the work
- Job interference by the owner or architect
- Owner constraints on scheduling interfaces
- Late availability of owner-furnished materials
- Owner-imposed acceleration or deceleration
- Impractical or impossible milestones
- Extra work orders
- Contract change orders
- Schedule impacts
- Sequence of construction altered by owner or architect
- Changes in activity's start event
- Changes in activity's finish event
- Time extensions
- Utilization of scheduling float time
- Late schedule approvals

If a project schedule is not available during the production, any consistent scheduling method that illustrates the delay will help support the claim. Any change in the project schedule, imposed by either the owner or the architect, can become a basis for a claim by the prime contractor or activity subcontractor. The best way to prevent a claim is to utilize scheduling techniques. This form of scheduling documentation offers the greatest protection. As the project progresses, time extensions may be granted by the owner, or by the architect with owner approval. These would result in changes in the project schedule and could result in revised interface dates or changed completion dates. The contractor will probably keep a close watch on the status of the schedules that it submits to the owner or architect for approval. Departures or deviations in the contractor's schedule, from the schedule originally intended by the owner, may be considered as contract amendments after approval by the owner. If a contractor submits a shorter schedule than originally called for under the contract, and it is approved by the owner or the architect, any liquidated damages provided under the contract may now be applied to the early finish date shown in the shortened schedule. There is no question that the only way for a contractor to make a profit on a project is to get in and get out in the shortest possible time, thus reducing the home office overhead and labor costs allocated to that job.

Constructive Changes

A constructive change is an informal act authorizing or directing a modification to the contract, caused by the owner or architect through an act or a failure to act. In contrast to the mutually recognized need for a change, certain acts or failures to act by the owner, that increase the contractor's costs and/or time of performance may be considered grounds for a change order. This is termed a constructive change. However, it must be claimed in writing by the contractor within the time specified in the contract documents. Otherwise, the contractor may waive its rights to collect. Types of constructive changes include:

- Defective plans and specifications
- Architect or engineer interpretation of documents
- Higher standard of performance than specified
- Improper inspection and rejection of work
- Change in the method of performance by owner
- Change in the production sequence by owner
- Owner nondisclosure of pertinent facts
- Impracticability or impossibility of performance
- Changed conditions

Sometimes these are referred to as "unforeseen conditions" or "differing site conditions." The term "changed conditions" is typically used in all federal contracts. There is also a growing trend by many public agencies and a few private owners to adopt similar wording in their contracts. Failure of an owner to provide payment for changed conditions places the contractor in a difficult position. If the owner takes a hard-line position on this issue, the contractor may find it necessary to seek relief from the court, a process that is both lengthy and costly to both parties.

The court's policy is to make adjustments in time and/or price where unknown subsurface or latent conditions at the project site are encountered by the contractor. The purpose is to have the owner accept certain risks and thus reduce the large contingency amounts in bids to cover such unknown conditions. The federal government and many local agencies include provisions in their construction contracts that will grant a price increase and/or time extension to a contractor who has encountered subsurface or latent conditions.

Under the legal guidelines for changed conditions, an existing underground pipeline that was either not shown on the plans at all or was incorrectly located on the contract drawings would qualify as a changed condition. Unusually severe weather conditions for the time of year and location of the project may also qualify. The discovery of expansive clays in the excavation areas, if not accounted for on the geological site survey and not detected in prior soil investigations, also qualify as changed conditions. Severe rains or similar weather that prevent work from being done, or which in any way delay the project, may not always be excusable delays, and in some cases have been ruled by the court to be excusable only and not compensable.

Change Order Pricing

One of the most common causes of contractor claims occurs during attempts to price change orders. Typically, owner change orders contain a waiver clause that requires the contractor to guarantee that the price and time named in each individual change order represents the total cost to the owner for that change, and the contractor waives any rights to impact costs.

This, unfortunately, leaves the contractor only one recourse—the claims process. The court has traditionally held that an owner may not force a contractor to "forward price" its impact costs when pricing an owner-directed change as this is an uncontracted cost element to the contractor. The court has written: "While it might be good contract administration on the part of the owner to attempt to resolve matters relating to a contract modification (change order) during the negotiation of the modification, use of a clause which imposes an obligation on the contractor to submit a price breakdown required to cover all work involved in the modification cannot be used to deprive the contractor from its right to file claims."

Schedule Conflicts Generating Change Orders

This is an often misunderstood area in contractor claims against the project. However, the probability of recovery by the contractor as the direct result of such conflicts is good, insofar as the settlement is limited to the cost difference between the project cost as the plans and specifications are interpreted by the owner or architect and the contractor. Most construction contracts are called *contracts of adhesion,* which is a term applied to contract documents that are drawn by one party and offered to the other party on a take-it-or-leave-it basis, where there can be no discussion of terms nor contract modification by the other party. The contractor, however, does have one advantage. In case of ambiguity, the court will interpret the contract in the contractor's favor. This does not relieve the contractor from the obligation of building the work in accordance with the interpretation of the architect or design professional, but only assures that the contractor be paid.

Frequently, the contractor will find that outdated standards are specified, or that products are named that no longer exist that will obviously conflict with scheduled installations. Often, the specification will contain references stating that, wherever codes or commercial standards are specified, the contractor is obligated to use the latest issue of that standard existing at the time the project went to bid. Unfortunately, in many cases the fact is that the design was based upon an old standard that existed in a designer's files, or a standard that was current during the design phase, but which later may have been updated by the sponsoring agency without the designer's knowledge. Occasionally, serious difficulties arise from such practices, and the contractor has the legal right to assess the project cost difference resulting from the error. Perfect specifications are hard to find, so the contractor must make a reasonable interpretation at the time of bidding the job and will then be in a good position for recovery if a variance exists. As in all matters of contract documents, plan for the best and provide contingency planning for the worst. The contractor's interpretation must be based on what a reasonable person would interpret the documents to mean; then the contractor will have the court on its side.

Other problems that often come up in project scheduling are directly related to the conduct of the owner's representatives on the job, such as the architect, project manager, or site supervisor issuing changes in the work that are of such magnitude as to constitute a cardinal change which creates a breach of contract. Or, on a lesser scale but indirectly related, where one activity subcontractor may negligently delay another, which may result in the owner's seeking recovery from the negligent contractor to pay the contractor that was planned. Some of the types of scheduling problems that fall into this category include the following:

- Damage to work by other prime contractors in fast-track scheduling
- Breach of contract
- Cardinal changes
- Beneficial use by the owner of the entire project before completion
- Work beyond contract scope
- Partial utilization of the project before completion
- Owner nondisclosure of site-related information
- Owner's failure to make payments when due

Sophisticated scheduling techniques, such as CPM or other network scheduling systems and their associated computer-generated printout sorts, have also compounded problems between the project team and the work contractors. Often the owner is swamped with stacks of computer printouts that he or she either does not want or does not know how to read. Yet somewhere in that stack of paper may well be the key defense against claims being made against the project and the owner. The question of scheduling methods of operation and the impact of deviating from the anticipated schedule require a careful analysis to determine the reasonableness of the originally anticipated schedule in conjunction with the planned methods of operation.

The analysis of every schedule-related problem is unique, and every such analysis should include a review of the anticipated sequence and schedule together with a review of the actual progress of the work. The review must include an analysis of any delays and impacts caused by all parties to the progress of the project. It is in the presentation of the project schedule sorts, containing cost data and their supporting documentation, that most construction claims are the weakest. A carefully detailed reconstruction of all construction activities and their related costs must be made and presented so that they can be clearly identified for possible use in negotiating affirmatively or defensively. Construction-related organizations are reported to be poor in their cost recordkeeping systems (or lack of systems), and they traditionally lack ability to assign costs to possible claims areas.

An accounting firm is not the answer either, because often those cost items relating to potential claims are not identified by an accounting firm. Most accounting firms are not familiar with the cost-control procedures for construction contracts. A contractor who realizes this, and takes steps to structure and run cost-tracking and audit trail sorts, will provide his or her company with the best form of claims management. Remember, records are the contractor's first line of defense in construction litigation documentation if not by the accountant, then by the home office. The responsibility is ultimately assigned to the prime contractor for computer recordkeeping and tracking. The accounting operations must cover all the necessary payroll functions, and the following employment laws must have tracking:

- All wages earned by any person working for an activity subcontractor for hourly wages on the project are due and payable at least twice during each calendar month. Labor performed between the 1st and 15th days shall be paid for between the 16th and 26th day of the month. Work performed on the project between the 16th and the last day of the month must be paid between the 1st and 10th day of the following month. However, the salaries of administrative or salaried management employees may be paid monthly.
- In a case of dispute over wages, the activity subcontractor must pay all wages conceded by him to be due, leaving the employee the right to dispute the remaining amount.
- Every activity subcontractor must keep posted conspicuously at its place of work a notice specifying the regular pay days and time of payment.
- In the event of a labor strike, the unpaid wages earned by the striking employees shall become due and payable on the next regular payday.

Change Order Check List

This section contains detailed information on all of the following:

- Where change orders are generated
- How they are discovered
- Where to find hidden changes and items maximize total compensation
- Whose responsibility they are

- How to consolidate time, cost, and liability components of the change order
- Evaluate and quantify total project schedule impact
- Maximize prices for extra work
- Prepare a complete and forceful change order proposal
- Protect all your rights to get paid for all consequential effect, even after receiving "final payment"
- Use tested and proven negotiation strategies and tactics
- Recordkeeping for maximum protection
- Effectively prepare and present claims for damages
- Effectively interpret contract documents
- How to maximize advantages in arbitration and litigation

A. PREDESIGN

1. Adjacent Properties
   a. Have all properties adjacent to the site perimeter been reviewed in detail?
   b. Are there:
      Seasonal watercourses?
      Heavy traffic patterns?
      Other independent construction activities?
      Other
2. Boring (Subsurface) Data
   a. Are boring depths inconsistent?
   b. Are boring locations erratic or unusual?
   c. Are boring locations relevant to construction:
      Are borings provided outside the area?
      Are gaps left within the building area?
   d. What time of year were the borings taken?
3. Building Code Compliance
   a. Have any violations of the building codes been observed by any building official when the building permit was applied for?
   b. Do any portions of the design appear out of the ordinary?
      Headroom?
      Entrances/exits?
      Handicap provisions?
      Fire separations?
      Lighting? Ventilation?
      Other

4. Easements/Rights of Way
   a. Are there designated easements?
   b. If so, will they adversely affect your operation?
   c. Do local traffic patterns restrict access?
   d. Are there parking areas, traffic patterns, business, etc., at the contract limit line that will restrict operations in any way?
   e. If the answer to 4.a is yes, do you know all conditions?
   f. If a restriction to your operation is evident, has your estimate accommodated it in some way?
   g. If the answer to 4.f is no, should a reasonable prebid site investigation disclose the condition?
5. Inland Wetland Approvals
   a. Does any portion of the site encroach on inland wetlands?
   b. If so, are all appropriate approvals in place?
   c. If required approvals are not apparent, have you requested the confirming information from the owner?
6. Interference of Utilities Not Properly Shown
   a. Have the characteristics of all existing utilities been verified with each respective company?
   b. Has each company representative reviewed the details with you at the site?
   c. Is anything different from that represented on the plans?
   d. Are the current utility charges for the various tie-ins the same as those given at the time of bid?
7. Plan Approvals (Building Permit)
   a. Has the building permit been applied for at the earliest possible time?
   b. Were there any problems?
   c. Were any notes or corrections made on the plans?
   d. Has the permit been delayed in any way?
   e. Is a permit required and a Certificate of Occupancy?
8. Temporary Utilities - Availability Within Contract
   a. Necessary for temporary field offices?
9. Limit Lines
   a. Have you confirmed the anticipated conditions at the time of bid?
   b. Are conditions adequate?
   c. Are site conditions now different?
   d. Are additional telephone/power poles needed?
   e. Is power available at all (without generating equipment)?
   f. Is previously anticipated use of existing facilities now prevented?

g. Is temporary heat and protection now required due to owner-caused delay?
h. Is water available in sufficient amounts for construction?

B. THE CONTRACT AND BID DOCUMENTS

1. Award Date
   a. Has an extension in the contract award date been requested?
   b. If so, is there any basis upon which to ask for an increase in the contract sum?
   c. Will acceleration be necessary?
   d. Will a portion of the project now be placed into winter conditions as a result of the start-up delay?
   e. Do you have the strength to now require more favorable contract terms:
      Is your bid substantially lower than the next bidder's?
      Can you complete the project in less time than your competitors?
   f. Were you involved in design development?
   g. Is the owner tied to you in any way?
2. Named Subcontracts
   a. Are there owner-selected subcontracts on the project?
   b. Does any disclaimer exist that limits the owner's liability for subcontractor selection?
   c. Are the subcontract agreements themselves owner defined?
   d. Is any specific procedure in place to resolve disputes between two owner-defined subcontracts?
   e. Will the owner in fact make decisions (or will there be constant attempts to drop the responsibility on the general contractor)?
3. Price/Bid Allowances
   a. Are there allowances anywhere in the contract?
   b. If so, have all allowance items been bid or rebid yet?
   c. Will allowance items be awarded in time to prevent schedule interruption?
4. Contract Time
   a. Did the first schedule draft drastically exceed the allowed contract time?
   b. Did subsequent schedule drafts incorporate unusual or excessive compressions and accelerations?
   c. Did any long-lead items purchases dramatically exceed the originally anticipated deliveries?
   d. If so, were they for specified items?

e. Had the contract award date been extended?
f. Had the site start date been extended for an owner-caused reason?
g. If the answer to either 4.d or 4.e is yes, was the schedule logic affected?
h. Did extra work result?
i. Can clear cause-effect relationships be demonstrated to justify more contract time?

C. PLANS AND SPECIFICATIONS

1. Ambiguities
   a. Are notes without specific references common (such as "As Indicated," "See Specs," "See Plans," etc.?
   b. Have you taken the time to research each one to confirm that completing details do in fact exist?
   c. If so, have you discovered incomplete, conflicting, or missing references?
   d. If so, have you cataloged each instance for individual consideration?
2. Ceiling/Underfloor Spaces Conflicts
   a. Is there a contract clause clearly noting the subcontractors to be responsible for coordination of their work?
   b. Have all areas of potential conflict in the ceilings been properly coordinated:
      Is there enough room to pitch all pipe?
      Do pitched lines miss all rafters, steel, and concrete beams?
      Can all ducts/pipes pass below beams at all locations shown?
      Too many items occupy the same space in any area? If so, can enough space be made, or can anything be moved?
      Are there large ducts shown to cross large beams and/or other significant obstructions?
      Will all light fixtures fit in the remaining spaces?
      Are there elaborate architectural, structural, or special shapes continuing into the ceiling?
      If so, do other building systems or equipment penetrate any part of them?
      If so, have you confirmed the actual size of everything?
4. Changed Existing Conditions
   a. Has the estimate been reviewed for:
      All sitework considerations?
      Any interferences with existing structures?
      Any noted conditions of existing structures?

Locations, extent, makeup, and conditions of existing utilities?
Traffic patterns and site access?
Anticipated storage and staging areas?
Parking and security arrangements?

b. Have the estimators involved met with you at the site to review all items?

c. Have any changes between conditions existing now and those existing at the time of bid become apparent?

5. Column and Beam Locations

a. Have the structural drawings been reviewed in detail:
Are column layouts erratic or unusual?
Are there any unusually long spans requiring relatively large structural members?
Are there unusual shapes, angles, slopes, or connections?
Are elevation changes strained or confusing?
Are beam sizes all different (with different ceiling/underfloor spaces below them)?
Have the locations of all large beams been reviewed?
Are there unusual designs?
If so, is enough information included for proper shop drawing preparation the first time around?

b. After reviewing the architectural, plumbing, HVAC, and electrical plans:
Are listed column line dimensions between all designs consistent?
Are there large ducts/pipes shown crossing large beams?
Are there light fixtures in the areas of large ducts?
Does the sprinkler main cross large beams, ducts, pipes, or fixtures?
Do random spotchecks of architectural dimension strings reveal any discrepancies?

6. Design Change Indications

a. Are there a large number of apparent last-minute design changes? Are there:
Different styles of type or handwriting in the specifications?
Incomplete erasures?
Out-of-sequence reference marks or inserted pages in the specifications?
Different handwriting on the plans?
Different use of language for the same or similar remarks?

7. Design Discipline Interfaces

a. Has any review to this point revealed any problems at the points where design disciplines cross each other?

8. Duplications of Design
   a. Have any duplications been observed? If so:
      Is each description complete?
      Are the descriptions in different specification sections with different contractors involved?
      Are the duplications included in the same specification?
      Is the same work specified twice?
      Is different work specified for the same function?
      Are any of the available options preferred?
   c. In a review of relevant contracts, plans, and specifications:
      Are any or all contracts of an adhesion format?
      Are any subcontractors owner selected?
      Are the affected subcontracts "per plans and specs?"
      Are there modifications to any contract?
      Are the rules of precedence outlined in the specification?
      Are all affected plans noted to be the responsibility of the affected subcontractor(s)?
      Do the descriptions of work included in the affected and related specification sections help your case?
   d. Objectively analyze each duplication:
      Have all the reasons why each subcontractor should and should not have carried the work in their bids been considered?
      Should any contractor aware of the work have reasonably construed it to be included by another trade?
      Did anyone request clarification from the owner prior to bid?
      If so, is the request and/or response documented?
      Is each duplication clear and complete in itself?
   e. Is there a preferred solution:
      Does any solution involve your own time or money?
      Are the dollar estimates of each solution a consideration?
      Is the timing of any solution particularly good or bad?
      Is any potentially affected contractor more inclined to accept the extra work?
      Does any solution make more sense?
   f. Do grounds exist to convince the owner that duplicated work is in fact not included anywhere?
9. "Fat" Specifications
   a. Does a review of the documents reveal an unusually fat "front end?"

Extensive duplication in the general provisions?
Long and/or labored descriptions and instructions?
"Catch-all" phrases and boilerplate not specifically applying to project conditions?

10. Finish Schedule vs. Specification Index
    a. In a comparison of the Finish Schedule to the Specification Index:
       Is each item accounted for?
       Is each item included only once?
11. Inadequate Level of Detail/Missing Details
    a. If enough design information has not been originally provided:
       Will the architect respond now with the complete information?
       Is it confirmed in writing?
       Are there any additional cost implications?
12. Light Fixture Locations
    a. In overlaying the lighting plans on the reflected ceiling plans, are there conflicts in:
       Ceiling light fixtures?
       Emergency lights?
       Soffit lights?
       Exit lights?
       Undercabinet lights?
    b. In overlaying the architectural plans, are there conflicts in walls, soffits, or cabinets?
    c. In overlaying the HVAC plans:
       Are there conflicts in register, grille, and diffuser locations?
       Are equipment actual sizes accommodated?
       Does everything miss the lights?
    d. In overlaying the fire sprinkler plans:
       Do the heads miss the lights?
       Do the heads fall in the center or quarter center of the ceiling tile?
       Is there an architectural pattern in the ceiling tile that will change location preference?
    e. In overlaying the electrical plans:
       Do the smoke detectors miss the lights (and everything else)?
13. Match Lines and Plan Orientations
    a. Are match lines present?
    b. If so: Are they necessary?
       Are they in the same location every time?

Do they include the same information?
Is anything missing?
Are they complete and to the same extent on every plan?

c. Is the north arrow in the same place on each drawing?
d. Are the orientations the same for each plan?

14. Mechanical, Electrical, and N.I.C. Equipment
    a. Are differences highlighted in all approval submissions?
    b. Has a Letter to Subcontractors Regarding Equipment Coordination been sent?
    c. Has a Letter to the Architect Regarding Contract Equipment Coordination been sent?
    d. Has a Letter to the Owner Regarding N.I.C. Equipment Coordination been sent?
15. Numerous Details and Dimension Strings
    a. Have repeated designs been observed?
    b. Are there many instances of multiple dimension strings?
    c. If so, have spotchecks uncovered errors?
16. Performance and Procedure Specifications
    a. Are there any instances in which both the performance and procedure specifications occur for the same item?
    b. If so:

       Are they mutually exclusive?
       Can they be made to be compatible?
       Is one or the other more expensive?
       Is one preferred over the other?
       Has one been included in the Schedule of Values?
       Is it cost-prohibitive to accomplish both?
       Is time or material availability a factor?
       Is one more complete or otherwise more appropriate?
    c. Is one preferred over the other?
    d. Have all the details and arguments supporting your position been assembled?
17. Proprietary Restrictions (Public)
    a. Does the specification being considered:

       Name fewer than three acceptable suppliers?
       Include the words "or equal?"
    b. Do you intend to use an "equal" product?
    c. If so, does the owner want a credit change order?

d. If so, have you considered a Letter to the Owner Regarding Equal to Proprietary Item?
e. Has the owner rejected your "equal" submission?
If so, have you considered a Letter to the Owner Regarding Rejection of Equal to Proprietary Item?

18. Specification Section "Scopes"
 a. Does the design coordination process appear to have been done correctly?
 b. Are specific cross-references included?
 c. Does the Scope section appear to be complete?

D. SITE

1. Grades, Elevations, and Contours
 a. Has the entire site been photographed before any work has begun?
 b. Have the existing grades been spot checked for accuracy?
 c. If so, have any discrepancies been discovered?
 d. If so, has a detailed check been arranged?
 e. Have the locations of existing telephone, water, sewer, fuel tanks and lines, and gas lines been verified?
 f. Have the manholes been opened to spot check actual pipe inverts?
 g. Have the locations of telephone poles, street signs, pole guys, and any other constructions been checked to avoid interference with site improvements?
 h. Have the actual horizontal distances among telephone poles, light poles, manholes, drainage structures, etc., been checked for accuracy?
 i. Have any discrepancies discovered been documented in the most accurate and unquestionable manner available?

Material Costs

Material costs are neither wholesale, or retail prices. They are estimates of what most contractors who buy in moderate volume will pay as of the middle of the current fiscal year. Add delivery expense to the material cost for other than local delivery of reasonably large quantities. Cost of delivery varies with the distance from source of supply, method of transportation, and quantity to be delivered. But most materials dealers absorb the delivery cost on local delivery (5 to 15 miles) of larger quantities to good customers. Add the expense of any delivery when it is a significant part of the material cost. Remember to add sales tax when sales tax will be charged to the contractor when buying the materials.

Waste and coverage loss is included in the installed material cost. The cost of many materials per unit after installation is greater than the purchase price for the same unit of waste, shrinkage, or coverage loss during installation. for example, about 120 square feet of nominal 1" X 4" square-edge boards will be needed to cover 100 square feet of floor or wall. There is no coverage loss with plywood sheathing, but waste due to cutting and fitting will average about 6%. Costs in the materials estimate assume normal waste and coverage loss. Small and irregular jobs may require a greater waste allowance. Materials priced without installation (with no labor cost) include an allowance for waste.

Labor Costs

Labor costs for installing the material or doing the work described are listed in a separate breakout. The labor cost per unit is the labor cost per hour multiplied by the man-hours per unit. To find the units of work done per man in an 8-hour day, divide 8 by the man-hours per unit. To find the units done by a crew in an 8-hour day, multiply the units per man per 8-day by the number of crew members. Man-hours include all productive labor normally associated with installing the materials described. This will usually include tasks such as:

- Unloading and storing construction materials, tools, and equipment on site
- Moving tools and equipment from a storage area or truck on site at the beginning of the day
- Returning tools and equipment to a storage area or truck on site at the end of the day
- Normal time lost for work breaks
- Planning and discussing the work to be performed
- Normal handling, measuring, cutting, and fitting
- Keeping a record of time spent and work done
- Regular cleanup of construction debris
- Infrequent correction or repairs required because of faulty installation
- Required safety meetings

Labor cost includes the basic wage, workers' compensation insurance, the employer's contribution to welfare, pension, vacation, apprentice funds, plus all payroll tax and insurance based on wages. Hourly labor costs in commercial construction are higher than those used in residential construction, reflecting the fact that craftsmen on industrial and commercial jobs are often paid more than craftsmen on residential jobs.

Adjust the labor cost to the job you are figuring when your actual hourly labor cost is known or can be estimated. Also adjust for unusual labor productivity. Costs estimates for labor are for normal conditions: experienced craftsmen working on reasonably well-planned and managed new construction with fair to good productivity. Labor estimates assume that materials are standard grade, appropriate tools are on hand, work done by other craftsmen is adequate, layout and installation are relatively uncomplicated, and working conditions don't slow progress. Labor costs will be higher where these conditions do not apply. Only experience and careful analysis can help you predict extras in the cost of unusual labor productivity. Supervision expense to the general contractor is not included in the labor cost. The cost of supervision and non-productive labor varies widely from job to job. Calculate the cost of supervision and non-productive labor relative to your project and add this to your estimate.

Equipment Costs

Expenses for major equipment (such as cranes and tractors) are grouped under equipment costs. Costs for small tools and expendable supplies (such as saw blades and tape) are usually considered overhead expense and do not appear in the equipment cost estimate. Equipment costs are based on rental rates and assume that the equipment can be used productively for an entire 8-hour day. Add the cost of moving equipment on and off the site. Allow for unproductive time when equipment can't be used for the full rental period. For example, the equipment costs per unit of work completed will be higher when a tractor is used for 4 hours during a day and sits idle for the remaining 4 hours. Generally, an 8-hour day is the minimum rental period for most equipment.

Subcontracted Work

Many sections in residential estimating require subcontracted work activities. Where this occurs, the cost includes material, labor, and equipment. However, subcontract costs also includes the subcontractor's supervision expense, contingency allowance, overhead, and profit. Costs in a residential contract that aren't identified as "subcontract" assume the general contractor does the work with his or her own crews. Work done by the general contractor's crews may cost more or less than the same work done by a subcontractor. A specialty subcontractor may be able to perform work at a lower cost, even after adding for overhead and profit.

Markup

The general contractor's markup is not included in cost breakouts. The national markup average for general contractors handling residential construction is 20 to 100%, but this will vary according to prevailing markets in your region, and must be area-adjusted. Apply this markup or some figure you select to all costs, including both subcontract items and work done by your own crews. To realize a markup of 20% on the contract price, you'll have to add 25% to your costs. Markup is the most difficult item to estimate.

A subcontractor's markup is added to both labor and materials costs. A markup of 30% is a mean average added to labor. Most materials costs include a 15% markup, but lower markups are typically used on major items of equipment which involve little labor cost.

Chapter 8

Dispute Resolution

Dispute Resolution

Dispute resolution should be a contractual obligation appearing within the specifications. The order of precedence of resolution steps should be mediation, arbitration and litigation, in that order. Arbitration should be further specified as binding arbitration. In this manner, precedence is already established for resolving the inevitable arguments that will occur throughout the project's production. The breakdown on each is as follows:

Mediation

Mediation means that all parties try to iron it out amongst themselves, with no outside consultants or lawyers becoming involved. If the contract is written with the above precedence for dispute resolution, and if the dispute cannot be resolved at the mediation level, all parties must proceed to arbitration before proceeding into litigation.

Arbitration

Construction differences that have gone as far as they can are then normally subjected to one of two different dispute resolution options—arbitration or litigation. If you have a choice between the two, it can indeed be a difficult decision. There are important differences between them, and your awareness of the advantages and disadvantages of each will allow you to determine which you will prefer for the specific case that you're considering.

The general guidelines that follow will help in determining which option will best suit your case. In any event, you are strongly urged to consult with a competent construction attorney before your final selection of the modus operandi.

Arbitration proceedings are much more informal than court litigation proceedings. Arbitrations are generally held in conference rooms in private locations. Arbitration is usually heard before a retired judge or member of the bar association. Arbitration is a lot less expensive than going to court. It is also a means to a quicker solution.

I have always advised my clients to include a binding arbitration clause in all contracts. If this clause is included, all parties are forced to arbitrate. The downside to arbitration is that, although arbitration is cheaper and quicker than court litigation, if your side loses it is almost impossible to appeal the judgment decision in arbitration. Proceedings are arranged and conducted by the American Arbitration Association. However, the actual proceedings are left to the arbitrator's discretion. Meeting considerations, such as time, location, and agenda, are worked out among both parties and the arbitrator.

Arbitration is typically faster than a court suit. The reasons arbitration is generally faster than a court action are because the filing procedure, arbitrator selection process, and calendar arrangements can proceed in as little as a month, although two to three months to complete the process is more common. This is dramatically faster than the court system, which may require literally years before a case can even find its way onto a court calendar.

The arbitration itself will generally require one to six months. This historical fact places it at approximately one-third the time required for a court proceeding. From these considerations, much of the literature on the subject goes on to debate the reasons why a shorter time frame for dispute resolution will be more "advantageous."

If you're a businessperson with any concern at all over cash flow, however, these discussions can be quite elementary. The fact is, as a businessperson, you must realize that anything you can do to speed up the decision/award will in all but the most unusual cases be well worth the trouble to implement. Arbitration speed, therefore, is a primary consideration in any dispute resolution decision. Arbitration expenses can sometimes be substantial, although typically less so than litigation. In arbitration cases lasting only one or two days, the arbitrator charges no fee. On the third day, the arbitrator begins to charge a fee that will be comparable to other legal professionals in the area (which will usually average three digits an hour), and proceedings will begin to get costly. Additionally, typically there is a filing fee for arbitration that is based on the settlement size of the claim. This filing fee is usually recoverable from the losing party by the winning party. These fees can be split between both parties as part of the final award, at the discretion of the arbitrator.

Arbitration will not be a dispute resolution option unless there is a specific agreement between the parties to arbitrate. This agreement can be in the form of an existing clause in the contract under which you're working or a separate agreement between the disputing parties drawn up at the time of the dispute resolution. On balance, the second situation is very difficult to accomplish if you are the one to determine that the case lends itself best to arbitration. Your opponent will likely see the disadvantage from his or her perspective. A subsequent agreement to arbitrate will then become very unlikely.

Arbitration proceedings are much more informal than is taking the dispute to court. Arbitrations are generally held in a conference room at any of a number of locations. Room arrangements, accommodations, and the proceedings themselves are arranged initially by the American Arbitration Association, but the proceedings are left to the arbitrator's discretion. All meeting considerations such as location, time, agenda, and so on can be worked out at the convenience and availability of the parties and the arbitrator(s) to a great extent, if the arbitrator(s) so allow. Depending upon the arbitrators' dispositions, the proceedings may follow either structured or less-disciplined styles.

There will be an arbitration panel of either one or three arbitrators. These people will have various construction-related backgrounds. They might be attorneys, contractors, architects, engineers, or anyone else with a familiarity with at least a portion of the construction business. They may be less familiar with the technical points of law, but their construction-industry savvy can help to keep issues on relevant tracks. Arbitration expense can be substantial, although not in every case. On small, relatively simple cases requiring only one or two days in arbitration, the arbitrator's time is at no cost. The arbitrator's fees, however, will begin on the third day of arbitration. They are established by the arbitrators themselves, and while arbitrators are not normally abusive in this regard, their fees will generally be set at levels roughly on a par with local fee structures of other competent professionals. In other words, they can add up rather quickly. In addition, there is a filing fee in arbitration that is based upon the size of the claim. In large actions, it can become quite expensive. If it is any consolation, the cost of the arbitrator and the fees are theoretically recoverable by the prevailing party. They can, however, be split between the parties as part of the final award. This again is solely at the discretion of the arbitrator. If the arbitrator is swayed very strongly for or against the merits of either position, it may affect the manner in which the arbitration expenses ultimately get charged.

Arbitration proceedings are not open to the public. The privacy of the parties is protected, and none of the documents is made part of the court record. If for any reason a public disclosure would concern you, this may be an important factor to consider.

Discovery

There is no "discovery" in arbitration. Discovery is a court process by which the opposition has the right (and can therefore be certainly expected) to go into your records and pull out every piece of information that can have the remotest possibility of bearing any relationship at all to the case. Likewise, you can (and probably will) do the same with your opponent's records. This process is designed around the principle of no surprises.

You can ask yourself if you really need to know in advance every single piece of information relating to the opposition's defense, statements, the remarks that their witnesses are going to make, and so on. If you can begin to imagine the time that it would take for each attorney to sift through each file, and multiply that time by his or her hourly rate, it shouldn't require a calculator to make the point of just how expensive and time-consuming the process can be. The fact is, many contractors and owners have elaborate paper trails surrounding any significant dispute. In many cases, therefore, the lack of the long, tedious, expensive discovery process leaves neither party at a real disadvantage. A complete review of the documentation that is already spread around will clarify to a very reasonable degree all positions that your opponent is going to take. If surprises do come up during the proceedings, your attorney will know the procedure for properly handling them on an item-by-item basis.

Evidence

The rules of evidence are much more relaxed in arbitration than in open court litigation. The requirements for the introduction of evidence in arbitration are much less stringent than is the court's, and their admissibility is again a matter of the judgment and discretion of the arbitrators. The admission of evidence by arbitrators has been historically a sensitive issue. This is because the failure of an arbitrator to consider an important item can be a basis for appeal. Therefore, even if an item is not admitted by the arbitrator as a formal exhibit, it can often be introduced by being "marked for identification." While this semantics game may satisfy some attorneys, the fact is it still allows items to be brought to the arbitrator's attention for his or her consideration.

Litigation

If there is no agreement to arbitrate, either established in the contract or subsequently arranged between the parties, there is no option to litigation. You must use the conventional legal system for dispute resolution. These are formal proceedings, as every court action is formal and rigidly structured. There is an elaborate set of rules that permits only technically correct claims to be filed. The technical merits of the case may play a disproportionate part in the proceedings and significantly increase the time necessary to present the case and/or defense. Like every other consideration of this section, this can work for or against you, depending upon the circumstances of your particular case.

Judges and juries, known in legalese as the *trier of fact*, will be deciding the merits of your case. It will be rare if either one has any construction experience, but you can count on the fact that both will be long on common sense. A serious disadvantage may be in their ability to grasp and believe important complicated concepts, such as off-site overhead expenses in a delay situation. Their ability to see through the fake may, however, far outweigh the disadvantage.

Although the court fees themselves are generally much less than the arbitration filing and arbitrator's fees, the other factors involving technical completeness and total time can increase total costs substantially. Only your experienced attorney will be able to lay out for you the details of where you can expect the total costs to wind up. Once you finally get on a court calendar, the actual time to proceed through the entire action may be between twelve months to two years. It is many times longer than arbitration, and can therefore lead to many times the total expense.

Court complaints are matters of public filings. They are open to the public and all documents are made a part of the court record. The proceedings themselves are open to anyone who cares to drop in. If this for any reason presents a problem, it will have to be well thought out before you proceed.

Discovery

The discovery process described under "Arbitration" is designed to prevent surprises on either party. However, in litigation, it is a process that is part of every significant litigation and can result in significant expense. It may be necessary for you, however, if your case is complicated and dependent upon information that is not already included in your own documentation.

Evidence

The strict rules of evidence in court are there to (in theory) protect a naive jury from considering items that do not have relevance to the case. "Naive" is not a criticism; it only describes the jury's lack of construction experience. You can expect their intelligence to make up for their lack of experience. The net effect, however, is that it is very time-consuming to introduce and explain each new piece of evidence. The most basic assumptions and construction business idiosyncrasies must be described in fundamental detail. Each piece of evidence must be evaluated against its admissibility criteria and debated. Technically questionable items may never see any opportunity to be presented at *all.* If these materials are important to your case, it can indeed have a detrimental effect. As with each other consideration, however, the effect can work for you if it is working against your opponent.

These guidelines have been intended to provide a general awareness of some major considerations when evaluating the different dispute resolution options. Beyond this, you are strongly urged to discuss these and other considerations with an experienced construction attorney before making any final decision on which way to proceed with your particular case. This specialist attorney is extremely important because anyone who tries to get into a game in which he or she doesn't know the rules and can't understand the procedures may be playing with less than a full deck. All commonsense understandings of the issues aside, the contract upon which you've supposedly based your actions is riddled with both obvious and very subtle language. While your experiences and familiarities may make the implications of certain provisions very clear to you, it is the hidden, yet extremely elaborate, meanings that very definitely place formal dispute resolution procedures outside the realm of the layperson. Beyond the contract documents' language and legal interpretations themselves, there are very definite rules and formalities that must be observed with absolute precision to even bring a case to its first step. Only a competent, experienced legal professional who specializes in construction litigation will be the person best suited to complete the arrangements and proceed with the action successfully.

In general, attorneys fall into several categories. The first distinction is litigating versus nonlitigating attorneys. Nonlitigating attorneys are not fighters. They are will writers, estate arrangers, real estate researchers, form filers, and audit managers. They are there to make sure that i's are dotted, t's are crossed, the paperwork is researched thoroughly, and documents are filled out properly. They fulfill extremely important functions, and their role in our society is indeed critical. While their services as far as we are concerned here may be necessary in some support kind of role, it is the litigating attorney in whom we're interested to take the lead in developing strategy and following through on the tough negotiations and ultimate court battles.

Expert Witnesses

The last section outlined important considerations in selecting the individual attorney who will lead the team through the ultimate dispute resolution option. Consultants who serve as expert witnesses can also play important roles not only in arbitration or litigation actions, but in each negotiation stage along the way as well. Even before these kinds of considerations, the efficient use of consultants' services in a preventive kind of approach to project management—before or during construction—may be their best use after all. Effective analysis and convincing presentations of cause, effect, and consequences may indeed prevent costly legal actions.

During legal actions, their presentations can clarify an otherwise complicated and confusing chronology of the detailed project record. The stories, arguments, and counterpoints that have been spread out over a period of months can decisively be tied together in a meaningful, clear presentation that will be well received and ultimately depended upon by confused juries and arbitration panels. The major areas in which consultants as expert witnesses may be extremely useful in bolstering your case are in conflicting testimony concerning:

- Project management
- Construction scheduling
- Claims/delay analysis

- Technical design analysis
- Cost controls
- Productivity improvement

Good expert witnesses can uncover the real facts by conducting efficient, detailed reviews of all relevant project documents. They can reconstruct events, and then be able to demonstrate all resulting consequential effects. They should be familiar with client-attorney relationships and be able to understand their role between them. They can aid your counsel in understanding the technical aspects of an issue, so that your own lawyer can prepare around it most effectively. In this capacity, the consultant can also aid your lawyer in document review to the extent that the relevance of each piece of information can be most accurately assessed. Good consultants will be able to prepare and present a variety of impressive graphic devices to deliver points with precision and power.

On the other side (in a defense or rebuttal situation), it will be the competent consultant who will best be able to refute technical evidence or any related presentation by the other side. If your opponent, for example, presents a detailed, complicated construction schedule to show the effects of a delay, find a good schedule consultant who will be able to find all the flaws in it (and fast) so your attorney will be able to chip away quickly at the schedule's legitimacy and destroy the presentation.

Fees are determined by the consultant's caliber and reputation, but also by the scope and extent of the work involved for your particular case. The extent of research necessary, the sessions required to coordinate their work with your attorney, and the labor hours necessary in preparation of demonstrative evidence, presentations, and testimony will vary greatly. Fees, therefore, can really only be established after the consultant knows more about the extent of involvement with these kinds of details. Take the time to review all the details of the consultant's involvement with your attorney *before* the services are contracted for.

Know exactly what you can expect to receive as a direct benefit, and insist upon estimates of total expenses. Discuss with your lawyer the merits of the extra expense as weighed against the lack of using such a service. If you make the decision to use the consultant, monitor all early operations very closely. Be certain that everyone understands your concern for cost efficiency as related to the consultant's fee, and make them aware that you're paying close attention to the outputs.

Litigation Claims Checklist

Because of the dramatic variety in extent and composition of construction claims, the manner in which each may be presented, the disposition of your own office and that of your attorney, and a variety of other factors, an absolute, comprehensive checklist that will cover all situations is really not practical. The detailed consideration for your particular case, under your own unique set of circumstances, can only be organized absolutely by your construction litigation lawyer. The construction claims checklist presented is this book is therefore designed as a good beginning. It is a starting point that will first provide a good understanding of the kinds of documents and information that your attorney will require and of the categories of people who may become necessary along the way. It is a checklist for you to review with your lawyer to determine what and who is or is not appropriate for your particular problem__this specific time. Once you're through considering the prearranged categories, continue the process during the same meeting to expand the list of remaining items and people that you will need. When you're done, you'll be left with a good understanding of the real level of involvement of all those named, and you'll begin to be able to see the effect that their use in the dispute action will have on the rest of your operation. Everything listed on the checklist will not be necessary for every claim. Their inclusion on the form will, however, continually bring them to your attention in each new situation. If they do not apply, simply cross them off.

Typical Case Histories

An owner contracted with a licensed professional contractor and, using the contractor's project schedule, constructed a $1 million custom home exactly per approved plans, only to find out during inspection that the structure did not meet code compliance due to an error by the building inspector. Who was liable? This case scenario appears in many stories about planning or building departments being liable for mistakes they have caused or overlooked. Here are some facts based on previous court cases. Let's first deal with building officials. In several cases, such as Chaplis v. County of Monterey, the court held that, while the building official issued a permit in violation of a county ordinance, the county and its employees were immune from liability. However, when the trial court in a construction-defect case dismissed a complaint against a building department official in Mammoth Lakes, the appellate court reversed the decision and held that the plaintiff stated a good cause if the representations made by the building official were shown to be false, or malice can be proved on the part of the building inspector.

So you see, it could go either way. The court has traditionally held that building departments and their employees are immune from liability, and that errors in design that are carried through the construction process revert to the architect or design professionals who are held to be liable, as they are supposed to be experts and are held to a "higher level of responsibility and accountability," and should be informed enough to spot an error before or during construction.

A contractor on a $5 million multiresidential construction project submitted a CPM project schedule that had been meticulously prepared. Its court deposition stated that the only objective of the CPM schedule was to get the project done as quickly and cost-effectively as possible. Due to a public works department-caused delay in utilities hookup, the contractor fell behind schedule. Later, the contractor submitted a claim for additional money and a 45-day extension.

The owner's architects analyzed the contractor's network and showed that contractor's succeeding activity event on the next major milestone on the critical path was delayed only 20 days. The awarding authority contended that 20-day delay was all the contractor was entitled to. The contractor stated it was only through its diligence in accelerating its work that the delay was held to only 20 days, and the contractor should be entitled to the full 45 days. The final court judgment was that the subcontractor could not support its 45-day claim because of improper schedule monitoring. The owner's architects had been able to do a better job of network analysis than had the prime contractor, and the contractor had to settle for the expenses it could prove. Despite the fact that it was the public works' delay, the contractor was, in effect, financially penalized for maintaining the schedule because of inadequate schedule monitoring by the contractor.

In Miller v. Delcon Development Corporation, Mr. Miller was a subcontractor who complained to the project owner, Delcon Corp., that the prime contractor had failed to pay him for his time and materials. The owner wrote a letter to Mr. Miller requesting that Miller send them a stop notice. Mr. Miller never did, but he did file a preliminary notice in a timely manner. The court held that if a subcontractor in a construction contract does not respond to a letter such as the one written by the owner, the subcontractor loses its lien rights and stop notice rights. The final judgment was that the subcontractor had lost his lien rights as a rights as a result of not giving the owner a stop notice. His only recourse was to sue for claims-liquidated damages in civil court.

Multiple Critical Path

A prime contractor on a $14 million multiresidential building project submitted a project schedule in compliance with the owner's specifications. There were four zero-float paths and seven critical paths. The prime contractor later stated that he was anticipating that the project team would delay one or more of the seven activities that were a part of the critical paths, thus allowing the contractor to file claims against the owner, which the contractor did at project completion. This type of bid getting is known in the trades as "low-balling." During mediation, the owner's project team cited the seven critical paths as an unrealistic approach. The contractor responded that it was the contractor's plan of work and the owner did not have any right to alter it once the contract had been signed. The mediation went unresolved into arbitration and later into litigation.

What had the contractor done to his network, and what was achieved? First, he increased the duration of routine concrete pours to three times their normal duration. If requested to shorten this duration, he planned to request payment for accelerated work. To prevent this, the owner had, as stated in the contract specifications, the right to use CPM and other standard planning tools, such as estimating guidebooks, to determine reasonable durations of all activities. Exceptions were to be negotiated between them both. The contractor used "policy or management constraints" (preferred way of doing work) to consume float. These constraints prevent activities from starting until preceding activities are sufficiently complete. On detailed analysis, this appeared illogical; however, the contractor was asked to apply workaround techniques, and he countered by saying that the changes to his plan of work would result in extra work, and therefore more claims against the owner. An owner should have the right to demand that constraints of this type be removed, and the work be replanned when the constraints threatened the critical path. Difficulties such as those just described might be overcome with adequate specifications. However, one can hardly hope to anticipate all the problems that can arise, particularly if the architect or, more likely, assistant specifications writer has a limited knowledge of all the tricks that can be used in CPM scheduling.

The specifications should protect both the owner and the contractor. But it should be remembered that the owner pays for the scheduling system and is entitled to get what he pays for. The specifications should set forth restrictions on falsifying networks to eliminate float. They should spell out what rights the owner has to utilize float to his advantage. They should clarify areas where the owner has the right to apply standard estimating techniques to activities for which contractors have obviously set over-long durations to eliminate float.

Equitable Adjustment

Whether the work is proceeding on a time and materials basis or the change has not been finalized for any other reason, precise documentation is the rule that must not be compromised. All direct-cost items must be kept track of starting with the effort that went into the proposal preparation. These historical records may become important if it later becomes necessary to demonstrate just how much disproportionate time the change has been taking away from other important project activities. Whether or not the change has been finalized, it is important to realize that these factors result from direct costs only. Others, termed "equitable adjustment" cover different costs.

Equitable adjustment determines that a contractor is entitled to receive compensation for all costs that would not have been experienced had the change not occurred. The obvious costs to be tracked includes items such as schedule impact (including job logic changes and milestone start delays), material invoices, payroll records, long-distance phone calls, loss of production efficiency due to outside effects (such as winter conditions), excessive management time devoted to research, increased surety and bonding costs, and delay to contract payments that cause excessive carrying costs. If these are not completely documented along with their respective justifications, they will not get paid for. Time and materials (or cost-plus) is usually provided for very clearly in the contract's change clause. AIA Document A201-1976, Article 12.1.4, provides for the cost-plus alternative. It states that cost-plus pricing will be the accepted standard, "If none of the methods set forth . . . is agreed upon." Most other change clauses will contain something similar, outlining T&M or cost-plus as the final method to determine costs—only after all more definite cost-establishing methods fail. For the contractor, T&M minimizes risk. The owner, however, perceives the method (and sometimes rightfully so) as a blank check.

The concept of equitable adjustment is one of the most important factors in any basis for settlement. To equitably adjust a contract, the contractor must be left in the position that he or she would have been in had the change not occurred. Equitable adjustment includes:

- All standard and legal costs
- Office and site overhead
- Interest
- Cost of special meetings
- Travel
- Special studies

Change Clause

Beyond the rules of general contract interpretation, the construction industry has developed a wide variety of distinct contract clauses that are unique to the industry and require special consideration. These clauses are applied consistently throughout the majority of modern construction contracts. The first of these is called the *change clause*. The change clause authorizes the owner to alter the work if the change falls within the general scope of the original contract. Without this express provision written into the agreement, there is no assumed right of the owner to make changes that the contractor will be obligated to perform. By including the change clause, the contract will now give the owner that right. The contractor, therefore, must be aware of the need for a countering defense mechanism in the contract. This takes the form of an efficient *dispute clause* that specifically details the procedure agreed to by all parties for dispute resolution. Most change clauses include the following provisions:

- No changed work is to be performed without a properly executed change order. (This provision is often expanded to exclude those instances where a contractor must act in an emergency to prevent injury or damage.)
- Any adjustments to the conditions of the contract may only be effected by a change order.
- The change order must be in writing, signed by both the owner and the architect.
- The change order must be for work that falls within the scope of the original contract.

- The change order must specify both the adjustment in contract price and net effect on the project schedule.

Commonly encountered construction claims arising from the application of change orders can be typically traced back to disagreement between owner and contractor on whether the change does in fact fall within the scope of the original contract, and if the incorporation of the change order will have the net effect of reducing the allowed percentage of the items to be billed on contract work. In typical change order clauses the owner, without invalidating the contract, may order changes in the work within the general scope of the original contract consisting of additions, deletions, or other revisions, the contract sum, and the contract time line being adjusted accordingly. All such change orders work must be performed under the applicable conditions of the contract documents.

Pass-through Clause

The pass-through clause is one of the important provisions to a prime contractor in a subcontract agreement. It incorporates into a subcontract by reference all or parts of the rights and responsibilities of the prime contract to the owner as they relate to the perspective subcontract. The standard subcontract and trade contract forms of both the Associated General Contractors (AGC) and the American Institute of Architects (AIA) contain a form of the pass-through clause. It binds a subcontractor to the same rights and responsibilities that the prime contractor has to the owner that apply to the work of the respective subcontractor.

Dispute Clause

The contractor aware of the need for a countering defense mechanism in the contract in this situation must insist on an efficient *dispute clause* that specifically details the procedure agreed to by all parties for dispute resolution. The dispute clause is the contract mechanism that provides the specific procedure for resolution of serious differences between the owner and contractor. It is often combined with the change clause to produce the final steps in the progressive change order negotiation procedure to ultimate settlement of dispute resolution. The purpose of the dispute clause is to settle differences inexpensively and speedily, which will benefit both parties. It may detail a progressive approach through a series of steps that must be followed, or it may simply outline the ultimate resolution option. The clause may stand by itself and be easily identified as follows from this example from AIA Document A201: "Any claim, dispute, or other matter in question. . . shall be subject to arbitration upon written demand of either party."

Reasonable Review

Contractors are responsible for reviewing all bid and contract documents as they relate to their work and submitting a proposal that is responsive to the requirements. What can be read, seen, or reasonably inferred gets broken out and priced. What has been intentionally or unintentionally left to the imagination does not find its way onto an estimate sheet in a competitive situation. Contractors are also responsible for disclosing obvious patent errors. The introductory provisions of the contract general conditions will normally require the contractor to visit the site prior to bid and accommodate all considerations that are readily apparent. The contractor is not responsible for performing a complete search of the contract documents to seek out hidden flaws in the contract, make subsurface explorations, or undertake any other costly investigation. This kind of detailed attention should have been given to the contract by the owner and designer.

In the scope of work, the intent of the contract is likely to come up at some point. Architects and owners will invariably use standard, but ambiguous, contract document language to detail the item in question in lieu of specifics. This is dangerous for the unwary contractor. For example, a common clause encountered is from AIA Document A201—1976, Article 1.2.3: "The intent of the Contract Documents is to include all items necessary for proper execution and completion of the work." Another classic is AIA Document A201—1976, Article 2.2.10: "All interpretations. . . shall be consistent with the intent of and reasonably inferable from the Contract Documents."

As you can see, these standard, but ambiguous, clauses do not define adequately the specific parameters of any specific construction job and often lead to contention and construction claims through many disagreements in performance contract gaps and an underlying effort to induce the contractor to perform additional work. The key to applicability (or lack of) design intent as a means for the owner or architect to fill gaps left in the contract is in the phrase "reasonably inferable." As with the duty to disclose patent errors and the responsibility for a "reasonable review," design intent will apply only if the gap being bridged is so obvious that a professional contractor would not normally overlook it.

Procedure Specifications vs. Performance Specifications

A procedure specification describes exactly the material to be used and its physical relationship to the remaining construction. Procedure specifications detail the quality of material, its composition and properties, as well as assembly of materials and components.

Procedure specifications must be comprehensive and correct in every detail to avoid ambiguity that will lead to construction claims. The accuracy and detail of the architect's design must be relied on by the contractor to achieve the intended construction result. If the procedure specification is followed exactly, the risk of design success lies completely with the architect or engineer.

A performance specification describes the ultimate function to be achieved, leaving the method and means of construction up to the contractor. Performance specifications specify performance characteristics on completion of a component or system, without indicating the precise materials to be used in the construction. The contractor is obligated to meet materials specifications designated in the stated design criteria.

Chapter 9

Construction Claims

RFIs, RFCs, and Milestones

A contractor should never attempt to construct any questionable area, or change, without first submitting to the owner or architect a request for clarification, or notifying them of an error. This is known as an RFI (request for information), or an RFC (request for clarification). In most contract forms, failure to do this may bar full recovery of contractor costs. These clarification documents must be in hand before the schedule's milestones become due. Milestones are benchmarks of long-term and short-term progress events that require some reasonable progress in activities or tasks toward a desirable outcome. Milestones are specific, time-limited objectives which, in combination, are sufficient to achieve those progress goals. Milestones are placed in the schedule to produce:

- Motivation for project team and contractors
- Utilization of competitive advantages
- Repair schedule competitive disadvantages
- Acceptable progress if those milestones are met on time and within budget
- Measurement of actual progress versus scheduled progress

At this stage of the process in checking RFIs and RFCs for meeting milestones, the contractor should be thinking in broad terms. As one studies particular areas of the schedule, one develops progress objectives that can be related as milestones. Milestones should reflect both strategic planning as well as current period objectives. They should also represent success. The remediation of existing problems will invariably lead to change orders and claims. This is an important concept for the professional contractor to grasp and use in developing the schedule. Unscrewing a problem isn't progress if the problem should not have happened, so milestones should mark progress only.

Positive benchmarking is important; milestones should be specific and measurable objectives. They should also appear as projections of progress evaluations at the end of each construction phase. Interim milestones should measure phases of progress which are critical to meeting long-term project timing. They should "back down" from the specified completion date to the present in sequential job logic order.

Once you have identified milestone phase areas, schedule dates for those events and the completion objectives that go with them. If you are scheduling with the owner's project team, it is important to agree upon the goals and objectives of each phase, leading to the establishment of milestones as benchmarks to evaluate the progress towards those goals and objectives. It is necessary to agree upon milestones prior to establishing phases, because milestones should lead the completion of each phase.

To reach milestone events on time, it is necessary to back down to the present. If you are to reach the critical completion date, where must the project be in one month? In two months? By end of quarter? To actually achieve your scheduled milestones, it will be necessary to periodically re-evaluate start and finish events, and use float to close gaps in timing. In planning the schedule, don't spend unreasonable amounts of time on making "perfect phases," but instead create obtainable and realistic milestones by which you can measure the project's progress.

The best planning process is to estimate durations based on what you know, and your company's historical data. Include in the schedule periodic sorts, or summary reports, that will test the milestones objectives for the coming quarter. This tactic will improve future milestone projections. Remember, the purpose of planning is to reduce risk. The best schedule plan is one that allows you to test your estimates before they cost lots of money. This is always your best perspective. We do this by running "what if?" scenarios with varying activity early and late starts, and varying the activity early and late finishes. A test of the schedule's quality is the extent to which completion events are balanced against float. To avoid construction claims, every project schedule needs to have the following criteria present:

- Each macroactivity step, defined into phases. This should be detailed enough to be measurable and give guidance to those who are responsible for managing that area of work.
- The person responsible for seeing that the activity is completed. This is usually the prime contractor, the site supervisor, a project team member, or an activity subcontractor.

- Activity cost in relationship to total budget. This should include time estimates, indirect, and direct costs. In the budget sort, a calculation needs to appear with this data to build a cost-tracking analysis.
- Project phases deadlines. These should include the actual time available by the subcontractor responsible for the activity, including its assigned early and late start and finish events.

All of these four steps should be closely monitored up to and including project close out. In larger and more sophisticated projects, this basic outline of plan elements will be more complex but will still follow these basics in a larger configuration. Time and cost estimates are made in greater detail for each activity. "Event dependencies" are established, which must be accomplished before others can be begun. Resources that must be devoted to each are listed, then broken out from the total budget. Resources are then allocated with priority weighing to ensure that critical path activities are resource-covered first. Progress-checking milestones and specific sorts are built into the procedure to ensure project schedule compliance.

Change Claims

Contractors using their own project scheduling can circumvent some of the general conditions and specifications of the contract by the use of claims. The contractor anticipates that the owner will delay at least two of the critical paths (on a multiple critical path network) that make up the project's production schedule, thus allowing the contractor to file claims against the owner. Claims then become the monetary reimbursements, usually in cash or canceling off certain punchlist items chargeable to the contractor, that are levied against the owner for delaying the contractor's work schedule. This always results in cost overruns blowing the budget and resulting in claims. The tactic of eliminating as much float as possible from the network puts the owner at a cost disadvantage in network analysis, and smart contractors know this.

So, the experienced owner may try to nullify this contractor's claims technique through clauses in the schedule specification that direct the contractor to redraw the network any time it is behind schedule on any specified critical path. In CPM project scheduling, this condition is instantly shown on the sort-by-total-float file in the negative float column whenever an activity falls behind schedule. The owner will attempt to resolve these and other scheduling problems and will always move to gain management control over claims generation. Usually, the owner or architect will make the stipulation in the contract for one or more of the several network scheduling techniques currently available. Such specifications would be written with the intent to boilerplate (cover all bases for) the owner's protection. With a combination of appropriate general conditions and supplemental general conditions, such specifications can be quite complex. One can easily see that the whole process can, and often does, wind up in claims and litigation. Many court decisions have been rendered on such claims, but none seems to have solved the problem. Now that you see both perspectives, you can understand how, as a prime contractor, your use (or nonuse) of different types of float in your phase scheduling can reduce (or generate) change orders and claims.

Shop Drawings

Shop drawings are detailed blueprint and schematic drawings that describe how a subcontractor plans to install and/or manufacture the project's equipment components that relate to their specific activity. They are prepared by the activities subcontractors and suppliers and are used in the contract to show how their installed equipment and work will meet specifications established by the structural engineers and the architect. Under the most common arrangement in current contract conditions, the prime contractor and the architect share the responsibility for shop drawings review. The initial shop drawings are normally reviewed by the prime contractor reveal if the equipment to be installed and work to be done will meet the design criteria. The prime contractor then stamps them with its approval and sends them for processing to the structural engineers and architect. These architect and structural engineers will then also review and approve the shop drawings and return them to the prime contractor. In doing so, the design professionals authorize the work to proceed in the manner detailed in the submitted shop drawings, or corrections needed to be resubmitted. It is at this point of contractor-designer reviews that areas of responsibility must be made clear to all parties. If these contractor and designer responsibilities are not spelled out in the contract conditions, the lack of responsibility will delay the shop drawing approval process. Each party (at this point) will be free to decide for itself how much responsibility it wishes to absorb. Obviously, the answer is as little as possible. Often, the subcontractors include less information than the designers require.

The design professionals then return the submitted shop drawings to the prime contractor for correction and withhold authorization to proceed pending additional shop drawings. The same condition exists with the design professionals in that they also wish to limit their professional liability and attempt to narrow the scope of their shop-drawings reviews in the contract. The combination of both parties shirking responsibility contributes to substandard shop drawings. This leads to increased confusion among all parties, resulting in exactly the opposite effect, the assumption of even more liability. The court has held that it cannot be argued that the contractual roles of the structural engineers or the architect are in an evolutionary phase created by changing contract structures. The legal precedent here is that the design professional's responses can compromise the thoroughness and attention to detail necessary for public safety in using the construction. Prime contractors often try to place much of the burden on the design professionals. This, in turn, invites confusion and substandard construction, ultimately leading to dissatisfaction at every step of the process. Historically, this effect increases liabilities for both the design professionals and the prime contractor.

The architect is normally responsible for approving shop drawings and reviewing them for their conformance with the original design concept. The difficulty lies in the fact that there seems to be no clear or consistent description of precisely what the designer's responsibilities in the shop drawing review and approval process include. To make matters worse, some design firms treat the shop drawing approval responsibility with less importance than it truly deserves. While it may be true that many designers do understand the critical nature of information incorporated in shop drawings, and treat them with respect and proper attention, some do not. Time and expense pressures in some design offices create a temptation to assign the shop drawing review task to a junior member of the staff, with lesser experience in such matters. In this case, the likelihood of errors in detecting deviations from conformance with the original design concept increases. This lack of clear definition and assumption of approval responsibilities has left many design offices, and their respective liability insurance companies, practically determining for themselves what their responsibilities will include. In addition, many architects and engineers have attempted to limit their liabilities by avoiding the use of the word "approved" in their shop drawing remarks altogether. Contract conditions phrasing written in the contract like "No Exceptions Taken," "Furnish as Submitted," "Examined," or the real cute one currently making the rounds, "Not Rejected," have now become the rule rather than the exception.

Moreover, the shop drawing stamps have been supplemented with elaborate language explaining what is and is not being done, in an effort to define the review process and minimize the design professional's liability.

There is legal opinion, however, that supports the idea that the designer's stamps may actually increase their liability, if their contract obligations for shop drawing review exceed the limits of the language included on the stamp. The owner and contractor rely on the designer's approval responsibilities as defined in the contract. If a designer operates in a narrower view as defined in the loose language on a stamp, it is a clear admission that the designer is doing less than he or she is contractually obligated to do. The court has upheld that *ex post facto* language in review stamps do not relieve public safety concerns in contractual obligations.

Beyond the basic designer/contractor approval responsibility clarification, another concern regarding shop drawings is designer response time. Common language in construction contracts notes that a designer will "review and approve shop drawings with reasonable promptness so as not to cause a delay in the work." "Reasonable response" is another ambiguous term that lacks precision. "Reasonable" becomes defined by trade practice in the project's geographic location. If nothing better is available in the existing contract language, establish the definition of "reasonable time" at the very first job meeting. Ten working days (except in unusual or adverse conditions) is standard. It's always best to pin it down early.

To emphasize the precise requirement, indicate on each submission transmittal the exact response date that approval of the respective item is required by before the project is impacted. A review of the specifications may establish the procedures for shop drawing submissions. Is there a special stamp required with the project name, and other identifying data that must be used on each shop drawing submission? If so, the stamps will have to be procured. How many copies of shop drawings are required for submission? How many copies of product data or equipment catalog sheets are required for submission? How are samples to be handled? Are all the types of shop drawings to be sent to the architect, or will structural drawings be sent directly to the engineer with an information copy only to the architect?

All these procedures should be clarified in the contract general conditions.

Long-Lead Items

Long-lead items are those items or materials within activities that are going to have to be manufactured, custom fabricated—or will take a long delivery time, such as out-of-area or out-of-state transport. The best time for subcontractors' ordering of long-lead items is after review of the plans and specifications.

Many subcontractors will wait until the last moment to order items of these types because they usually require partial down payment, and no one wants to have out-of-pocket expenses until necessary. However, if you allow subcontractors who have long-lead items in their activities to order those items only as needed, the subcontractor is running the risk of down time if the items are then unavailable or on back order.

Making sure long-lead items are ordered after going over the plans and specifications ensures timely delivery with no activity delay further setting back other critical path activities. We ensure that subcontractors have indeed done this by requiring that copies of their purchase orders to suppliers of long-lead items be given to the project manager within a specified time after the signing of contract. This requirement should appear in the contract specifications, and guarantees that items are ordered in advance. Attention must be given to the longest delivery items (long-lead items) on the schedule to see which ones fall on the critical path. If there are several large field-erected items with varying completion dates, you might even break the most important item into individual units to track the latest one. The same critical analysis applies to time-critical subcontracts. Long-lead items need purchase orders confirmed in advance by all associated subcontractors. Purchase order copies should be supplied by subcontractors and suppliers, in advance, to the prime contractor to maintain adequate lead time between ordering and confirming the order in the project schedule, along with delivery within critical event windows in that activity. When shop drawings are sent to the architect or engineer, a note should be entered on the accompanying transmittals, indicating whether any are to be expedited or if there is an order of importance.

The contractor must monitor the shop drawing log from time to time to make sure that the drawings are being cycled in a timely manner. The shop drawing log should show what action has been taken on each drawing when the drawing is returned. Although the transmittals forwarding the copies and the transmittals accompanying the returned drawings will be the documents of record, the shop drawing log will highlight the activity. The shop log should contain a column showing the new disposition of the drawings after they have been reviewed by the designers. If any drawings are disapproved, they would probably be returned to their originator. If approved, they will be distributed to their originator, with copies to any concerned subcontractors, a copy to the field, and one for the office file.

Shop Drawings Log

Shop drawings are detailed blueprint and schematic drawings that describe how a builder plans to install the project's equipment components that relate to specific production activities. They are prepared by the builder (as prime contractor), activities subcontractors, and suppliers and are used as contract documents to show how their installed components and equipment will meet specifications established by the design engineers and the architect. Under the most common arrangement in current contract conditions, the builder and the architect share the responsibility for shop-drawings review. The shop drawings reviewed by the architect reveal if the components and equipment to be installed will meet the design criteria. The architect then stamps them with his approval and sends them for processing to the structural engineer. These design professionals will then also review and approve the shop drawings and return them to the owner. In doing so, the design professionals authorize the work to proceed in the manner detailed in the submitted shop drawings, or corrections needed to be resubmitted. It is at this point of design review that areas of responsibility must be made clear to all parties. If these builder and designer responsibilities are not spelled out in the contract conditions, the lack of responsibility will delay the shop-drawing approval process. Each party will be free to decide for itself how much responsibility it wishes to absorb.

Obviously, human nature being what it is, the answer is as little as possible. Often, the subcontractors include less information than the designers require. The design professionals then return the submitted shop drawings to the builder for correction and withhold authorization to proceed pending additional shop drawings. The same condition exists at the other end of the process with the design professionals, in that they also wish to limit their professional liability and attempt to narrow the scope of their shop-drawings reviews in the contract. The combination of both parties shirking responsibility contributes to substandard shop-drawings that are not adequately detailed for modern equipment component manufacturers, suppliers, and distributors. This lack of responsibility leads to increased confusion among all parties, resulting in exactly the opposite effect, the assumption of even more liability. The court has held that it cannot be argued that the contractual roles of the structural engineers or the architect are in an "evolutionary phase" created by changing contract structures. The legal precedent established here is that the design professionals' responses can compromise the thoroughness and attention to detail necessary for public safety in using the construction. Builders often try to place much of the burden on the design professionals. This, in turn, invites confusion and substandard construction, ultimately leading to dissatisfaction at every step of the process. Historically, this effect increases construction claims for both the design professionals and the builder.

The architect is normally responsible for approving shop drawings of modular construction and reviewing them for their conformance with the original design concept. The difficulty lies in the fact that there seems to be no clear or consistent description of precisely what the designer's responsibilities in the shop-drawing review and approval process include. To make matters worse, some design firms treat the shop-drawing approval responsibility with less importance than it truly deserves. While it may be true that many designers do understand the critical nature of information incorporated in shop drawings, and treat them with respect and proper attention, some do not. Time and expense pressures in some design offices create a temptation to assign the shop-drawing review task to a junior member of the staff, with lesser experience in such matters.

In this case, the likelihood of errors in detecting deviations from conformance with the original design concept increases. This lack of clear definition and assumption of approval responsibilities has left many design offices, and their respective liability insurance companies, practically determining for themselves what their responsibilities will include. In addition, add in what was previously mentioned about many architects and engineers attempting to limit their liabilities by avoiding the use of the word "approved" in their shop drawing remarks altogether. The chance for error in project specifications increases. Moreover, the shop drawing stamps have been supplemented with elaborate language explaining what is and is not being done, in an effort to define the review process and minimize the design professionals' liability.

There is legal opinion, however, that supports the argument that the architect's or designer's stamps may actually increase their liability, if their contract obligations for shop drawing review exceed the limits of the language included on the stamp. The owner and builder rely on the designer's approval responsibilities as defined in the contract. If a designer operates in a narrower view as defined in the loose language on a stamp, it is a clear admission that the designer is doing less than he or she is contractually obligated to do. The court has upheld that *ex post facto* language in review stamps do not relieve public safety concerns in contractual obligations.

A shop-drawing log should be started so that when the shop drawings start arriving from subcontractors and suppliers, each drawing can be properly logged into the computer for tracking. Not only will a log of shop-drawing activity allow the prime contractor to keep track of what drawings have been received, it will also show where the drawings have been sent and how long they have been there. The prime contractor needs to prod each subcontractor and supplier to submit their drawings promptly. At the first job meeting, the major subcontractors and/or material suppliers should be requested to submit a preliminary shop-drawing submission schedule.

The log should include the major pieces of equipment for which shop drawings are required, and the anticipated date when each drawing will be submitted. The subcontractor and/or supplier should also include the approximate delivery date of equipment after the approved shop drawings have been returned to them. Once shop drawings have been received, the next hurdle is getting them approved in a timely manner.

The architect and the owner should review the incoming shop drawings before the prime contractor logs them into the master schedule, to determine whether they conform to the project specifications and requirements. If compliance is questionable, contact the party who made the submission to re-examine the shop drawings to verify contract specifications. If there are deviations, it might be best to note them before submitting the drawings to the architect.

It is at this point that the prime contractor must establish credibility with the architect and engineers and must show that the shop drawings are being reviewed for compliance with the plans and specifications and are not merely being passed through without any scrutiny whatsoever.

Beyond the basic designer/builder approval responsibility clarification, another builder's construction claim concern regarding shop drawings is designer response time. Common language in construction contracts notes that a designer will "review and approve shop drawings with reasonable promptness so as not to cause a delay in the work." "Reasonable response" is another term that lacks precision. "Reasonable" becomes defined by trade practice in the project's local standards. If nothing better is available in the existing contract language, establish the definition of "reasonable time" at the very first job meeting. Ten working days (except in unusual or adverse conditions) is considered standard. It's always best to pin it down early.

To emphasize the precise requirement, indicate on each submission transmittal the exact response date that approval of the respective item is required by before the project is impacted. A review of the specifications may establish the procedures for shop-drawing submissions. Is there a special stamp required with the project name, and other identifying data that must be used on each shop drawing submission? If so, the stamps will have to be procured. How many copies of shop drawings are required for submission? How many copies of product data or equipment catalog sheets are required for submission? How are samples to be handled? Are all the types of shop drawings to be sent to the architect, or will structural drawings be sent directly to the engineer with an information copy only to the architect? All these procedures should be clarified in the contract general conditions.

The log should contain a column showing the disposition of the drawings after they have been reviewed by the designers. If any drawings are disapproved, they would probably be returned to their originator. If approved, they will be distributed to their originator, with copies to any concerned component manufacturers, subcontractors, a copy to the field, and one for the office file. The log should also include the major pieces of equipment for which shop drawings are required, and the anticipated date when each drawing will be submitted. The manufacturer, subcontractor, or supplier should also include the approximate delivery date of equipment after the approved shop drawings have been returned to them.

Once shop drawings have been received, the next hurdle is getting them approved in a timely manner. The architect and owner should review the incoming shop drawings before the builder logs them into the computer, to determine whether they conform to the project specifications and requirements. If compliance is questionable, contact the party who made the submission to re-examine the shop drawings to verify contract specifications.

If there are deviations, it might be best to note them before submitting the drawings to the architect. It is at this point that the builder must establish credibility with the architect and engineers and must show that the shop drawings are being reviewed for compliance with the plans and specifications and are not merely being passed through without any scrutiny whatsoever.

Care must be taken by the prime contractor to discern which subcontractors should receive select informational copies of shop drawings. For instance, when a mechanical subcontractor is being sent an approved copy of its boiler shop drawing, the electrical contractor should have an informational copy in order to confirm the line voltage requirements. All too often a piece of equipment is ordered with electrical characteristics at variance with the voltage requirements shown in the drawings. If an error such as this can be caught in the shop drawing stage, there may be little or no additional cost involved to make the equipment compatible with the building's electrical system.

Long-Lead Items Purchase Orders

Attention must be given to the longest delivery components and prefabricated items (long-lead items) on the schedule to see which ones fall on the critical path. If there are several large field-erected components with varying completion dates, you might even break the most important components into individual units to track the latest one.

The same critical analysis applies to time-critical subcontracts. A typical shop drawing is used on all components and long-lead items. In addition, a separate file should be kept for each trade that will be submitting large numbers of drawings, such as structural steel, plumbing, HVAC, sprinkler system, and electrical work. There are several software programs on the market today that allow the builder to create a fax modem transmittal forwarding the shop drawing to the architect or owner and transferring this information automatically onto a shop drawing log.

The information inserted at the time of preparation is stored and transferred to the shop drawing log. Although the shop drawing log contains material from various subcontractors, separate computer files and summary report sorts can be created for individual trades.

Modular components and long-lead items need purchase orders confirmed in advance by all associated subcontractors. Purchase order copies should be supplied by subcontractors and suppliers, in advance, to the prime contractor to maintain adequate lead time between ordering and confirming the order in the production schedule along with delivery within critical event windows in that phase activity. When shop drawings are sent to the architect or owner, a note should be entered on the accompanying transmittals, indicating whether any are to be expedited or if there is an order of importance. The builder must monitor the shop-drawing log from time to time to make sure that the drawings are being cycled in a timely manner.

A log similar to the shop-drawings log should be started of long-lead items purchase orders so that when the shop drawings start arriving from component manufacturers, site work subcontractors, and suppliers, each purchase order can be properly logged into the computer for tracking. Not only will a log of components and long-lead items purchase orders allow a builder to keep track of what purchase orders have been received, it will also show where the purchase orders have been sent and how long they have been there. The builder needs to prod each subcontractor and supplier to submit their purchase orders promptly.

At the first job meeting, the major subcontractors and/or material suppliers should be requested to submit a preliminary purchase order submission schedule. The log of long-lead items purchase orders should show what action has been taken on each purchase order when the order is received. Although the transmittals forwarding the copies and the transmittals accompanying the received purchase orders will be the documents of record, the full shop drawing log will highlight the particular component or activity.

Transmittals

The previously examined shop-drawing log is used on all long-lead and fabricated items. In addition, a separate file should be kept for each trade that will be submitting large numbers of drawings, such as structural steel, plumbing, HVAC, sprinkler system, and electrical work.

There are several software programs on the market today that allow the project contractor to create a fax modem transmittal forwarding the shop drawing to the architect or engineer and transferring this information automatically onto a shop drawing log. The information inserted at the time of preparation is stored and transferred to the shop drawing log. Although the shop drawing log contains material from various subcontractors, the separate drawings and schedules must be logged into the master with no delay.

Chapter 10

Strategies For Claims Defense

Overview

Contract change orders do more than alter the scope of the construction work. They also disrupt the orderly sequences of the project schedule, change methods for work not otherwise directly addressed by the change, change the project schedule's job logic, alter prior coordination, and interfere with planned deliveries. The contractor should be left in the condition he or she otherwise would have been, had the change not occurred. The concept that a contractor be made "whole" in the event of a changed condition that affects the work is called *equitable adjustment*.

If an adhesion contract or its terms are subject to more than one reasonable interpretation, the contractor has the right to choose the interpretation. The reasons are generally the same as those for ambiguity resolution. This legal assessment is consistent with the concepts of a standard of interpretation. It also provides that if a reasonable interpretation is available, there is no duty to continue the search for other "reasonable interpretations" that may or may not exist. It does not have to be the most reasonable interpretation, or even more reasonable than that of whomever drafted it.

Standard of Interpretation: Reasonable Expectations

Contract interpretation must be understood by every contractor. Many construction documents contain words like "reasonable" or "acceptable" or other such words that appear on the surface to clarify levels of quality and performance expected in the construction process. In practical application, however, exactly the opposite is likely to be the case. Whenever any adjective is used to describe some measure of acceptability that is not specifically definable or quantifiable (such as "flat" or "hard" as opposed to "level within 1/8" in 10' in any direction" or "95% relative compaction") a flag should pop up in your mind that another gray area is sure to develop into muddy water if it is not clarified now and advantageously. The word "reasonable" has been and continues to be the epitaph on many contractors' profits and would-be change orders.

There are standard fundamental rules of interpretation that apply to all contracts, regardless of their industry or their subject. These contract law interpretation principles form the basis of initial application of contract law in assessing how conflicts in the contract language and format will be decided. In the absence of definitive resolutions within the contract, application of these rules is used to provide accurate assessments of the suitability of various interpretation options. The first of these fundamentals of contract law interpretation is called *reasonable expectations*. The objective of reasonable expectations is to protect the interests of all parties to the contract. The first step then is to define clearly what those reasonable expectations are. Which of the parties has a closer meaning and interpretation that most closely matches those reasonable expectations?

Constraints

Constraints are delays placed upon an activity's starting event by practical limitations, preceding activity delay, or the owner's option as a means of control over the network time line by using types of float as time manipulation of that event. They are used as factoring tools in time scale computations of the network schedule. The only claims management fundamental in constraint on a noncritical path activity is that it cannot be delayed longer than the available free float. If a noncritical path activity is constrained beyond its float period, it therefore reaches negative float and automatically becomes part of the critical path. As the project proceeds, some of the activities that were previously on the critical path will no longer be there. Suppose there is a three-day delay to framing. Originally, framing had one day of float. Now the schedule must be readjusted, because the total project has now been delayed two days. Framing has now become a critical path activity, and concrete has moved off the path.

Remember, constraints are those practical limitations that can influence the start of certain activities. The activity that involves the placing of reinforcing steel obviously cannot start until the steel is on the site. Therefore, the start of the activity of placing reinforcing steel is constrained by the time required to prepare and approve the necessary shop drawings, fabricate the steel, and deliver it to the jobsite. Strategically, you treat constraints the same as activities and to represent them as event durations on the network diagram. These types of constraints prevent activities from starting until other preceding activities are sufficiently complete. Policy constraints are those methods the contractor uses in the project's production to consume total float. Constraints are shown as dummy arrows on a project schedule as dashed lines with zero elapsed time.

Strategic Use of Float

The float of an activity represents the scheduling-timing leeway available to be committed for total project critical path duration. When total float is available, the earliest start of an activity can be delayed, its duration extended, or a combination of both can occur as long as the late finish time is not exceeded. It is the window of timing that is available for that activity if the activities preceding it are started as early as possible and the ones following it as late as possible. When an activity has free float time available, this extra time may be utilized to accelerate or constrain succeeding activities. Computerized scheduling uses float as a time manipulation tool, and the nomenclature and workings of float comprise essential knowledge of the project's prime contractor. The total float of an activity is the maximum time that its actual completion can go beyond its earliest finish time and not delay the entire project. If all the total float is used on one activity, a new critical path is created. The free float of an activity is the maximum time allotted by which its actual completion date can exceed its latest finish event without affecting either the overall project completion or the times of any subsequent activities. If a work item within an activity is delayed enough to consume an activity's free float, if it still completes on time, the activities following it are not affected and they can still start at their earliest start times.

Float is computed by certain factors determined by the network time, as well as the unit factors of the Sort By Total Float/Late Start. The sort provides data columns for each factor of float, which are:

- The activity's description
- The activity's i-j number for computer identification
- The activity's estimated duration versus remaining duration
- The percentage of activity completion
- The activity's actual start and finish

- The activity's early and late start
- The activity's early and late finish
- The activity's remaining free float
- The activity's available total float

The computer determines each activity's float by computations between i-j numbers of events and corresponding timeliness. In addition, the computer can determine the critical path in each phase, the completion time for critical activities, early start and finish dates for each activity, free float, and total float available. If we use the job logic of two sequential activities, such as placing reinforcement steel for a concrete slab and the succeeding activity of pouring the concrete, we can examine them using the factors in the report to calculate float. If these activities are not on the critical path and the difference between the early and the late start dates is a total of five days, the total float the rebar activity has is five days—the difference between the early and late start, or the early and late finish.

In float computations within the scheduling, each activity does not have total float individually, but all the activities in all of the phases share the same total float time made from their individual free float. Phases of activities have total float, and if it is consumed by one activity, float no longer exists in that phase for any activities. Not a good situation for any contractor, and one that the pros avoid like the plague.

Ambiguities Resolved Against the Drafter

The law assumes that those responsible for drafting the contract will provide for their own interests and will have reason to be conscious of uncertainties. The owner and architect, for example, have had more time with the contract documents than the contractor to make sure they are complete and correct. The documents will also be boilerplated for their interests rather than the contractor's. If, after developing and reviewing the contract, the originator(s) were unable to resolve ambiguities, it is unreasonable to expect a contractor to find flaws in a short period of time prior to bid. However, it is important to be aware that this rule of interpretation only applies in the case of adhesion contracts and not where the terms were the product of joint efforts of both parties to the contract. Adhesion contracts are those that are produced by one party and are presented to the other party on a "take-it-or-leave-it" basis, without reasonable opportunity to negotiate. This is the typical type of contract in construction. Most contractors can therefore be comfortable with the knowledge that this rule will apply in their application of the lump-sum bids submitted to the owner on complicated forms that refer to the plans and specifications as is. But for the contractor to recover costs and damages resulting from such ambiguity, it must be demonstrated that the interpretation was in fact relied upon.

The Four Cs of Contracts

A good working strategy for claims management involves understanding that contracts have a better likelihood of avoiding change orders if they contain what construction management professionals refer to as the "four Cs of contracts." These begin with the first "C" standing for the contract being *comprehensible*. The contract general conditions and specifications must make sense. The requirements of the specifications and of the detail drawings must be described in a manner consistent with the reasonable level of competence that can be expected from those who will be applying those requirements. They must clearly make sense in the general scheme of the project's construction process. They should avoid language that misleads, is ambiguous, or would be otherwise more sophisticated than the situation requires. If such misleading language is present, it typically is a prelude to later confusion that will lead to extra charges.

The second "C" stands for *comprehensive*. The descriptions and references of each contract provision and construction detail must be effectively correlated with each other, be broad enough to encompass all affected areas in a manner consistent with the design intent, and combine to give adequate attention to every portion of the project. Assumption that a particular description will apply without specific indication stated in the general conditions or specifications will invariably lead to change orders and extra work. Never assume anything in the performance of your contractual obligations.

The third "C" refers to the contract documents being *clear*. The terms of the contract general conditions and specifications must be unambiguous. The meanings and descriptions of the terms, components, and references must be clear and presented without confusion in wording.

The final "C" stands for *complete*. The descriptions and references of each contract item and provision must include all relevant details. Material specifications especially must be examined to ensure all physical parameters are described sufficiently and totally. Contract clauses must accommodate all pertinent considerations. All this must be accomplished without unnecessary redundancy or duplication, and without any gaps in logistics of the specifications. If it isn't complete, don't bid.

Chapter 11

Proven Strategies for Winning Claims

Perfecting Claims

In perfecting construction claims, it is imperative that legitimacy and credibility be maintained above all else. This applies to whatever strategy, tactic, or countermeasure that may be employed throughout the entire negotiation, dispute resolution, and possible litigation. The first concern is to be sure to research the contract documents and specifications thoroughly. Before any action is taken, determine if there is any specific procedure outlining the precise steps to be followed in the event of serious disagreement that leads to dispute resolution. Start with the Change Clause and then move to the Dispute Clause. If nothing becomes apparent, don't be satisfied until you've convinced yourself that it hasn't been provided for in some other way. If you finally do locate some such procedure, such as an appeal to a higher level or "board of contract review," then it should be followed in a strict, literal sense. Often, however, such a procedure will not be included on residential contracts. In this case, it may be wise to advise the owner in your first correspondence that your action is predicated on necessity, given that the contract is deficient in this regard.

Tactics

As soon as a contract change is recognized to have occurred, thoroughly research the contract documents to confirm the case. Pay particular attention to any notice requirements that may exist. Upon such confirmation, immediately prepare and submit a complete change order proposal. Take great care to ensure that it complies with all proposal requirements in every respect. Give the change order as much legitimacy as possible. Make the additional work seem obvious. Since the burden of proof of the owner's conduct is on the contractor, get the owner's conduct down in writing and included into the official project record. Remove at least that portion of risk.

We've seen that if a specification is subject to more than one reasonable interpretation, the contractor has the right to choose the interpretation. This is one of the strongest leverage points in change order dispute resolution, and should be known as such by every residential builder and contractor. Now let's take a detailed look at other contractors' leverage points in construction claims.

Latent Conditions (Defects)

Latent conditions are those that by the time of the bid are unknown to anyone involved in the process. Because the proper relevant information usually is unobtainable by reasonable means, the presence of latent conditions may be nobody's fault. It is for this reason that owner approval of such change orders can be straight-forward (if only in principle). The two most common types of latent conditions are subsurface site conditions that were undetected by testing or records research such as subsurface rock interfering with excavation or utility lines in locations different than those on record) and hidden conditions in an existing structure that could only have been discovered by demolition in an alteration project.

Defective Specifications

Defective specifications refer to contract documents containing errors or omissions. Whether done manually or on computer, most specifications begin with some existing general standard format. It might be some existing specification that the architect used for a past similar project, or it might be a commercially prepared standard general format such as the ones produced by the American Institute of Architects (AIA) or the Construction Standards Institute (CSI). Also, sometimes the specifications are taken directly from manufacturers' brochures, which often contain facsimiles of specification sections specifically designed to make it easy for an architect to copy and include in the project specifications.

The difficulty arises when several other manufacturers are listed as acceptable and the words "or equal" are included. The "equal" manufacturers will have different specifications. If the differences are slight, they will be ignored. If they are significant, you may be entitled to a change order if forced to provide the "specified" item at an increase in cost. In either case, your specification didn't start out from the ground up with just your project in mind. By using an "off-the-shelf" specification, inherent problems such as language that is not relative or does not apply, or important inclusions necessary for comprehensive descriptions being overlooked, often occur in the real world application of construction.

The design process is the architect's primary priority. It is only when the design is complete that the technical specifications and the contracts themselves are finally assembled. This makes the technical specifications the architect's secondary priority. As they are not an architect's primary priority, technical specifications are looked upon as an unfortunate, tedious necessity by the design professional. To simply use a broad, generic specification available "off the shelf" is a fast and easy way to get it over with. What makes the situation even worse is the fact that junior architects or even clerks may be enlisted to prepare certain portions of the specifications. It's bad enough that the specifications you're working with had originally been written for another project and made to "fit" your job. Add to this the ambiguous effort described and it is easy to see how the probability of errors and omissions increases.

Incomplete Design

The term incomplete design describes the failure of a designer to specify adequately all project components to the level necessary to allow the contractor to proceed with the work. This type of deficiency can appear anywhere in the contract documents. The simple examples can amount to neglecting portions of equipment requirements.

More common, however, is an incomplete design resulting from the failure to verify that supplementary information is in fact being provided by someone else. The net effect is that information is left out entirely. Notations like "see specs" instead of "see section 123.4," or "see structural plans" instead of "see 6/S5," are clear indicators that an assumption was made that somebody other than the designer provided the relevant information. They are an indication that those on the design end didn't do their homework and verify the facts. The assumed information is likely to be missing altogether.

A lack of coordination between design disciplines will cause this same incomplete design problem. If separate plans and specifications are developed by individual consultants, the designs and specs may be completed without full or realistic consideration for the other disciplines involved, and the plans wrapped up with unfulfilled assumptions that someone else is responsible for the completeness of details.

Specifications Inconsistencies and Impossibilities

Every contractor at one time or another has restated the age-old phrase, "It's easy to draw lines on a sheet of paper." An architect, or more probably his or her assistant, may be blindly following standard engineering rules, fitting together components without consideration for following related construction practices, or not properly coordinating an item with adjacent work. Typical resulting difficulties include illogical construction sequences (job logic not being followed), foundation wall configurations that don't allow removal of forms, anchors that are to be installed in accessible areas, equipment that can't be accessed after installation for maintenance, and equipment that can't use available access (it won't fit through the door). Ring a bell? Happened to you more than once? My point exactly.

Nondisclosure

Nondisclosure is unquestionably an unethical tactic if used intentionally by an owner or architect. It is the failure to inform the contractor of design or construction information that is significant to the completion of the work. If intentional, it is presumably because the inclusion of such information would lead the contractor to accommodate it with a corresponding increase in price. If unintentional, nondisclosure can be the result of a deficiency in the design effort or simply the failure of an owner to understand the significance of the information to the construction process. In either case, the contractor has been denied the opportunity to consider all the necessary information at the time of bid. In most nondisclosure cases, the court has held that compensation for the effects of nondisclosure once they do become known is justified.

Deliberate errors also serve the same purpose as nondisclosure, they misdirect and deceive. People add wrong, multiply wrong, leave out words, move decimal points and commas, make incorrect statements, or change clauses after agreements have been made. A very subtle word change can alter the context of entire meanings. If an opponent discovers such an error later, someone in that organization must have the courage to explain it to the management. They must explain why it wasn't caught in the first place. From the error maker's point of view, the penalty is low, but the potential rewards are high. There is nothing really to lose at this point because anybody can make a mistake.

Be on the lookout for such errors. Force yourself to check the arithmetic, read the fine print, and confirm your understandings. If you are the recipient of a deliberate error, get angry. Never assume everything is right. Be skeptical and double-check all data.

Credit Turnarounds

The moment any portion of the work is deleted from the contract, the word "credit" pops into the owner's and architect's mind. Suddenly, despite the fact that you've been coordinating the item all along, submitted shop-drawings, placed material orders, and so on, the owner asks you for a big price break. Consider them closely for what they are. Begin with the original value of the work. Catalog all the activities that have been performed in relation to it, along with all the effects of the deletion on all other components of the project. Take all the time that you need to be certain that the list is complete. Refer to a detailed discussion of the relevant considerations of component prices. Items with small original costs might actually cost the owner more money to delete them from the contract at this late stage. Larger items won't be the windfall credit that the owner is expecting. Know what the details are, and you won't find yourself giving the store away.

Deadlines

Always give specific dates for your change order proposal acceptances. Make them real. People don't believe deadlines until one has been uncomfortably imposed upon them. When the date of acceptance of your proposal runs out, immediately resubmit a revised proposal reflecting all appropriate cost increases. They'll begin to take your deadlines seriously. The deadline that you set will very likely be the earliest date of your response. In contrast, if you don't set a deadline, you set the stage for procrastinated action.

Deadlocked Negotiations

There is nothing wrong with negotiation deadlock in itself. If you are not apprehensive about deadlocks or the problems that they can create, you will be able to consistently demonstrate a firm attitude to your opponent. The concern, then, is how to break the impasse that you don't want. Consider using these strategies:

- Change the negotiator.
- Change the time shape of uncertainty. Postpone some difficult parts of the agreement for renegotiation at a later time.
- Change the shape of future satisfaction by offering additional guaranties.

Deadlocks put severe pressure on both organizations. If your opponents don't make the first move to break it, they might very well welcome yours. Reinvolve your opponent in discussions, and try to create a climate in which new alternatives can be developed.

Level of Detail

In representing change order components a large number of small items is better than a small number of large ones. Large items must be broken out in some manner anyway, if you really hope to explain the price to your opponent. Why make it more difficult than it needs to be? Individual small prices (and a lot of them) are easy to discuss, are readily demonstrated, and don't create much friction. The small dollars make it easier to give in to a persistent request. People don't like to appear petty (even if they are). Break apart your big items into as many small ones that you can come up with, and you'll consistently be able to justify higher prices for the total than you otherwise might have thought.

Chapter 12

Permit Responsibilities

Overview

A building permit is a license to proceed with the work but not authority to violate, cancel, alter, or set aside any of the provisions of the technical codes. Issuance of a building permit will not prevent the building department from thereafter requiring a correction of errors in plans, construction, or violations of this code. Claims for permit responsibilities are often generated from these additional corrections. The contractor must be aware that every building permit issued becomes invalid unless the work authorized by such permit is commenced within six months after its issuance, or if the work authorized by the permit is suspended or abandoned for a period of six months after the time the work is commenced. One or more extensions of time, for periods not more than 90 days each, may be allowed for the permit. The extension must be requested in writing and justifiable cause demonstrated. Extensions must be in writing by the building official.

Whenever a permit is issued in reliance upon an affidavit or whenever the work to be covered by a permit involves installation under conditions which, in the opinion of the building department, are hazardous or complex, the building official may require that the architect or engineer who signed the affidavit or prepared the drawings or computations supervise such work. In addition, they (the design disciplines) may be responsible for conformity with the permit, providing copies of inspection reports as inspections are performed, and upon completion make and file with the building department a written affidavit that the work has been done in conformity with the reviewed plans and with the structural provisions of the technical codes. In the event the architect or engineer is not available, the owner must employ in his stead a competent person or agency whose qualifications are reviewed by the building official.

When the building official issues a permit it is endorsed in writing or by stamp on both sets of plans as "Reviewed for Code Compliance." One set of reviewed drawings is retained by the building department and the other set is returned to the applicant. All construction work requiring a permit cannot start until the permit holder posts the permit card in a conspicuous place on the jobsite. The permit must be protected from the weather and located in such position as to permit the building inspector to conveniently make the required entries. This permit card must be maintained until the Certificate of Occupancy or Completion is issued by the building department. The permit drawings must be kept at the jobsite and be available to the building inspector.

A permit will not be issued until the fees prescribed by the local building department have been paid. Nor will an amendment to a permit be released until the additional fee, if any, due to an increase in the estimated cost of the building, structure, electrical, plumbing, mechanical, or gas systems has been paid. Any person who commences any work on a building, structure, electrical, gas, mechanical, or plumbing system before obtaining the necessary permits is subject to a penalty of 100% of the usual permit fee in addition to the required permit fees. The building department keeps a permanent and accurate accounting of all permit fees and other monies collected, the names of all persons upon whose account the same was paid, along with the date and amount.

On all buildings, structures, electrical, plumbing, mechanical, and gas systems or alterations requiring a permit, a fee for each permit must be paid as required at the time of filing application, in accordance with the schedule as established by the applicable governing authority. If, in the opinion of the building official, the valuation of building, alteration, structure, electrical, gas, mechanical or plumbing systems appears to be underestimated on the application, the permit will be denied, unless the applicant can show detailed estimates to meet the approval of the building official. Permit valuations must include total cost, such as electrical, gas, mechanical, plumbing equipment and other systems, including materials and labor.

Permit fees are set by the independent building department authorities that adopt the building codes in the area of jurisdiction where the project is located. With the exception of small storage outbuildings, fences, and painting, permits are required for all new construction, repairs, or alterations. Permit fees serve two purposes for the local jurisdiction. First, they are calculated to underwrite the expenses of the building department in general and administrative costs. This avoids putting the burden of the cost of the building department on the taxpayers of the community and puts the cost burden upon the project owners and developers of the construction projects within the jurisdiction.

The second purpose of permits is to initiate the timetable and scheduling of the required inspections. Additionally, a copy of the permit is forwarded to the county or city tax assessor, who will reassess the property valuation for an increased tax rate. In most jurisdictions, permits are required to be posted on the jobsite in a conspicuous place for accessibility of the building inspector. Any owner, authorized agent, or contractor who plans to construct, enlarge, alter, repair, move, demolish, or change the occupancy of a building or structure, or to erect, install, enlarge, alter, repair, remove, convert, or replace any electrical, gas, mechanical or plumbing system, the installation of which is regulated by the technical codes, or to cause any such work to be done, must first make application to the building official and obtain the required permit for the work. Permits are not required for the following mechanical work:

- Any portable heating appliance
- Any portable ventilation equipment
- Any portable cooling unit
- Any steam, hot, or chilled water piping within any heating or cooling regulated equipment
- Replacement of any part which does not alter its approval or make it unsafe
- Any portable evaporative cooler
- Any self-contained refrigeration system containing 10 lb (4.54 kg) or less of refrigerant and actuated by motors of one horsepower (746 W) or less.

A special building permit for a limited time must be obtained before the erection of temporary structures such as construction sheds, seats, canopies, tents, and fences used in construction work or for temporary purposes such as reviewing stands. Such structures must be completely removed upon the expiration of the time limit stated in the permit. A building, electrical, gas, mechanical, or plumbing permit carries with it the right to construct or install the work, provided the same are shown on the drawings and set forth in the specifications filed with the application for the permit. Where these are not shown on the drawings and covered by the specifications submitted with the application, separate permits are required. Ordinary minor repairs may be made with the approval of the building official without a permit, provided that such repairs do not violate any of the provisions of the technical codes. Each application for a permit, along with the required fee, must be filed with the building official and must contain a general description of the proposed work and its location. The application must be signed by the owner, or its authorized agent. The building permit application must indicate the proposed occupancy of all parts of the building and of that portion of the site or lot, if any, not covered by the building or structure. It must also contain such other information as may be required by the building department. An application for a permit for any proposed occupancy is deemed to have been abandoned six months after the date of filing for the permit, unless before then a permit has been issued. One or more extensions of time for periods of not more than 90 days each may be allowed by the building official for the application, provided the extension is requested in writing and justifiable cause is demonstrated.

Difficulty in the terms of the permit responsibility has generated claims concerning the difference between plans and specifications. Problems sometimes arise in conflicts between plans and specifications, therefore the builder must understand the relationship between the two. In construction law, approved plans, specifications, and general conditions have equal authority and none takes precedence (sequence or logical order of authority) over the others. As covered in Chapter 1, the working blueprints of a project are known as the plans. For normal residential construction, plans are broken down into six categories: plot plan, foundation plan, floor plan, roof plan, elevations, and detail cross-sectionals. Larger commercial and industrial projects will call for many additional architectural, structural, mechanical, plumbing and electrical plans. The plans typically encountered on modern high-end residential projects run about 200 pages. The only set of plans that are legal to build with are the stamped and approved set. These can be (and often are) revised prior to or during the construction process. Plans represent the proposed description of the construction process, including installation and materials to be used. Officially stamped plans represent the approved description of the project's construction process. Specifications, on the other hand, provide greater detail on facts regarding equipment, materials, and installation than are shown on the plans.

When required by the building official for permit processing, two or more copies of specifications, and of drawings drawn to scale with sufficient clarity and detail to indicate the nature and character of the work, must accompany the application for a permit. Such drawings and specifications must contain information, in the form of notes or otherwise, as to the quality of materials, where quality is essential to conformity with the technical codes. Such information must be specific, and the technical codes must not be cited as a whole or in part, nor must the term *legal* or its equivalent be used as a substitute for specific information. All information, drawings, specifications, and accompanying data must bear the name and signature of the person responsible for the design. The building official may require details, computations, stress diagrams, and other data necessary to describe the construction or installation and the basis of calculations. All drawings, specifications and accompanying data required by the building official to be prepared by an architect or engineer must be affixed with their official seal. The design professional must be an architect or engineer legally registered under the laws of the state regulating the practice of architecture or engineering and must affix his or her official seal to the drawings, specifications, and accompanying data, for the following:

- All Group A, E, and I Occupancies
- Buildings and structures three stories or more high
- Buildings and structures 5000 sq. ft (465 m^2) or more in area

For all other buildings and structures, the submittal must bear the certification of the applicant that some specific state law exception permits its preparation by a person not so registered. The only exception to this is that Group R3 buildings, regardless of size, require neither a registered architect or engineer, nor a certification that an architect or engineer is not required. (UBC and BOCA have different requirements on Occupancy Group R3. Check UBC Section 310.1, and BOCA Section 302.1.) Plans for all buildings must indicate how required structural and fire resistance integrity will be maintained where a penetration of a required fire-resistant wall, floor, or partition will be made for electrical, gas, mechanical, plumbing, and communication conduits, pipes, and systems and also indicate in sufficient detail how the fire integrity will be maintained where required fire-resistant floors intersect the exterior walls. Drawings must show the location of the proposed building or structure and of every existing building or structure on the site or lot. The building official may require a boundary line survey prepared by a qualified surveyor. The building department may also require the following of the permit applicant, be it owner or contractor: A general site plan drawn at a legible scale which includes, but is not limited to, the location of all buildings, exterior storage facilities, permanent access ways, evacuation routes, parking lots, internal roads, chemical loading areas, equipment cleaning areas, storm and sanitary sewer accesses, emergency equipment and adjacent property uses. A building floor plan drawn to a legible scale must be included that shows all facilities within the building and indicates rooms, doorways, corridors, exits, and fire-rated assemblies with their hourly ratings. The building department will examine each application for a permit and the related accompanying documents—consisting of drawings, specifications, computations, and additional data—and will determine whether the construction indicated and described is in accordance with the requirements of the technical codes and all other pertinent laws or ordinances. The building official may accept a sworn affidavit from a registered architect or engineer stating that the plans submitted conform to the technical codes. For buildings and structures, the affidavit must state that the plans conform to the laws as to egress, type of construction, and general arrangement and, if accompanied by drawings, show the structural design and that the plans and design conform to the requirements of the technical codes as to strength, stresses, strains, loads, and stability. The building official may accept such affidavit provided the architect or engineer who made the affidavit agrees to submit to the building official copies of inspection reports as inspections are performed and upon completion of the structure, electrical, gas, mechanical, or plumbing systems a certification that the structure, electrical, gas, mechanical, or plumbing system has been erected in accordance with the requirements of the technical codes. Where the building official relies upon such affidavit, the architect or engineer must assume full responsibility for the compliance with all provisions of the technical codes and other pertinent laws or ordinances.

Permit Delays

The building department tries to act upon an application for a permit without unreasonable or unnecessary delay. If the building official is satisfied that the work described in an application for a permit and the contract documents filed conform to the requirements of the technical codes and other pertinent laws and ordinances, they issue a permit to the applicant. If the application for a permit and the accompanying contract documents describing the work do not conform to the requirements of the technical codes or other pertinent laws or ordinances, the building department will not issue a permit, but will return the contract documents to the applicant with the refusal to issue such permit. Such refusal must, when requested, be in writing and must contain the reason for refusal.

When application for permit to erect or enlarge a building has been filed and pending issuance of such permit, the building official may, at his or her discretion, issue a special permit for the foundation only. The holder of such a special permit is proceeding at its own risk and without assurance that a permit for the remainder of the work will be granted nor that corrections will not be required in order to meet provisions of the technical codes.

Prime Contractor's Permit Responsibilities

It is the responsibility of every prime contractor who must make contracts for the installation or repairs of building, structure, electrical, gas, mechanical, or plumbing systems, for which a permit is required, to comply with state or local rules and regulations concerning licensing which the applicable governing authority may have adopted. A permit will not be given by the building department for the construction or alteration of any building where the building is to be changed to affect the exterior walls, bays, balconies, or other appendages or projections fronting on any street, alley, or public lane; or for the placing on any lot or premises of any building or structure removed from another lot or premises, unless the applicant has made application at the office of the director of public works for the lines of the public street on which it is proposed to build, erect, or locate that building.

Inspections

This section lists and defines the inspections required during the various construction phases before a final inspection can be called for. For most residential projects, there are five required inspections:

- **Foundation inspection**. This initial inspection is done after footings are excavated and foundation forms and reinforcement steel are installed. This inspection checks for footing depth and width, conditions of footing base and sidewalls, and proper sizing, amount, and condition of reinforcement steel (rebar).
- **Concrete slab and under-floor inspection**. This inspection is done after all under-floor and in-slab plumbing, conduit, service, and ancillary equipment is in place but before the concrete is poured. This area must remain open for this inspection, which includes no subfloor, and floor sheathing may be installed prior to this inspection.
- **Frame inspection**. This inspection is done after the exterior and interior walls are framed; fire stops, freeze blocks and lateral bracing are installed; electrical and plumbing are roughed-in; and all vents are installed.
- **Gypsum board, lath and plaster inspection**. This inspection is done after all gypsum board (sheetrock) or lath work is installed but before it is taped and textured or plaster is overlaid upon lath. Here the inspector is checking for proper fastener schedule (placement) and condition of installation.

- **Final inspection**. When the construction is complete and the finish grading is done, the final inspection clears the project for issuance of a certificate of occupancy. The building is now ready to occupy.

Should any of these required inspections fail, the inspector will issue a correction notice calling for correction of the specific work that failed inspection before scheduling a follow-up reinspection. Typically, reinspections are an extra charge to the owner or contractor. Depending on the project's complexity, there may be other special inspections called for by the building inspector. Before issuing a permit, the building official may examine any building, electrical, gas, mechanical, or plumbing systems for which an application has been received for a permit to enlarge, alter, repair, move, demolish, install, or change the occupancy. The inspector will also inspect all buildings, structures, electrical, gas, mechanical, and plumbing systems from time to time during and upon completion of the work for which a permit was issued. He or she must make a record of every such examination and inspection and of all violations of the technical codes.

When deemed necessary by the building official, an inspection of materials or assemblies at the point of manufacture or fabrication will be required. A record must be made of every such examination and inspection and of all violations of the technical codes. The building official may accept reports of private inspectors of recognized inspection services, provided that after investigation the building official is satisfied as to their qualifications and reliability. A certificate called for by any provision of the technical codes will not be based on such reports unless the same are in writing and certified by a responsible officer of such service. The building official will inspect at various intervals all construction or work for which a permit is required, and a final inspection will be made of every building, structure, electrical, gas, mechanical, or plumbing system upon completion, prior to the issuance of the certificate of completion.

The building official upon notification from the permit holder will make the following inspections and such other inspections as necessary, and must either release that portion of the construction or notify the permit holder or its agent of any violations which must be corrected in order to comply with the technical codes:

Building

- Foundation Inspection: To be made after trenches are excavated and forms erected.
- Frame Inspection: To be made after the roof, all framing, fireblocking, and bracing is in place, all concealing wiring, all pipes, chimneys, ducts and vents are complete.
- Final Inspection: To be made after the building is completed and ready for occupancy

Electrical

- Underground Inspection: To be made after trenches or ditches are excavated, conduit or cable installed, and before any backfill is put in place.
- Rough-in Inspection: To be made after the roof, framing, fireblocking and bracing is in place and prior to the installation of wall or ceiling membranes.
- Final Inspection: To be made after the building is complete, all required electrical fixtures are in place and properly connected or protected, and the structure is ready for occupancy.

Plumbing

- Underground Inspection: To be made after trenches or ditches are excavated, piping installed, and before any backfill is put in place.
- Rough-in Inspection: To be made after the roof, framing, fireblocking and bracing is in place and all soil, waste, and vent piping is complete, and prior to this installation of wall or ceiling membranes.
- Final Inspection: To be made after the building is complete, all plumbing fixtures are in place and properly connected, and the structure is ready for occupancy.

Mechanical

- Underground Inspection: To be made after trenches or ditches are excavated, underground duct and fuel piping installed, and before any backfill is put in place.
- Rough-in Inspection: To be made after the roof, framing, fireblocking, and bracing are in place and all ducting and other concealed components are complete, and prior to the installation of wall or ceiling membranes.

- Final Inspection: To be made after the building is complete, the mechanical system is in place and properly connected, and the structure is ready for occupancy.

Gas

- Rough Piping Inspection: To be made after all new piping authorized by the permit has been installed, and before any such piping has been covered or concealed or any fixtures or gas appliances have been connected.
- Final Piping Inspection: To be made after all piping authorized by the permit has been installed and after all portions which are to be concealed by plastering or otherwise have been so concealed, and before any fixtures or gas appliances have been connected. This inspection must include a pressure test.
- Final Inspection: To be made on all new gas work authorized by the permit and such portions of existing systems as may be affected by new work or any changes, to insure compliance with all the requirements of this code and to assure that the installation and construction of the gas system is in accordance with reviewed plans.

Work must not be done on any part of a building, structure, electrical, gas, mechanical, or plumbing system beyond the point indicated in each successive inspection without first obtaining a written release from the building official. Such written release must be given only after an inspection has been made of each successive step in the construction or installation as indicated by each of the foregoing three inspections. Reinforcing steel or structural frame work of any part of any building or structure must not be covered or concealed without first obtaining a release from the building official.

Project Close Out Certificates

A certificate of completion closes out the construction process of the project and a certificate of occupancy is mandated by all the national building codes before the structure can be used. These are the contract documents that bring the legal record of construction of a building to a finish and allow occupancy to begin. If both these certificates are not issued, technically the building is unoccupiable as it is not finished being built. This will cause conflict with all unpaid accounts of the project. Utility companies will not service the building until the certificate is issued. Contractors who do not follow through to this stage in the construction recordkeeping phase can face abandonment contractual breech and loss of license from the Registrar of Contractors. A new building cannot be occupied or a change made in the occupancy, nature, or use of a building or part of a building until after the building official has issued a certificate of occupancy. These certificates will not be issued until all required electrical, gas, mechanical, plumbing, and fire protection systems have been inspected for compliance with the technical codes and other applicable laws and ordinances and released by the building official. Upon satisfactory completion of construction of a building or structure and installation of electrical, gas, mechanical, and plumbing systems in accordance with the technical codes, reviewed plans, and specifications, and after the final inspection, the building official will issue a certificate of occupancy stating the nature of the occupancy permitted, the number of persons for each floor when limited by law, and the allowable load per square foot for each floor in accordance with the provisions of the code. A temporary/partial certificate of occupancy may be issued for a portion or portions of a building which may safely be occupied prior to final completion of the building.

Permit Variances

The Construction Board of Appeals may vary the application of any permit in any particular case when enforcement would be contrary to the spirit and purpose of the building code and finds all of the following:

- That special conditions and circumstances exist which are peculiar to the building, structure, or service system involved and which are not applicable to others.
- That the special conditions and circumstances do not result from the action or inaction of the applicant.
- That granting the variance requested will not confer on the applicant any special privilege that is denied by this code to other buildings, structures, or service system.
- That the variance granted is the minimum variance that will make possible the reasonable use of the building, structure, or service system.
- That the grant of the variance will be in harmony with the general intent and purpose of this code and will not be detrimental to the public health, safety, and general welfare.

In granting a variance, the board will probably prescribe a time limit within which the action must be commenced or completed or both.

"Approved" Materials

The term "approved" has proven to be a major source of claims contention between builders and owners. To clarify the situation, the head building official having local jurisdiction on any specific installation is the only person who can legally decide what materials and/or equipment are "approved." The label of a nationally recognized testing laboratory on a piece of equipment is a sure and ready way to be assured that the equipment is properly made and will function safely when used in accordance with the application data and limitations established by the testing organization. Each label used on a product gives the exact name of the type of equipment as it appears in the listing book of the testing organization.

It should be noted by the contractor that the definitions for "labeled" and "listed" do not always require that the listing laboratory be "nationally recognized." The code definitions acknowledge that a local inspector may accept the label or listing of a product by a testing organization that is qualified and capable even though it operates in a small area or section of the country and is not "nationally recognized."

Design Plans Checking

In scrutinizing the project's plans for probable change order indicators, start with verifying that code requirements are being met everywhere in the basic design. In the building code, residential occupancy mandates that all living, dining, and sleeping rooms are required to have windows that must open directly to the outside. This residential classification difference represents far less hazard to the public than other types of occupancies and the fire marshall is strict on it. The exception to this is that the windows may open onto a roofed porch if it has an inside ceiling height of at least 7 feet and is 65% open on the longer side. In design computations for number of windows required for each room, the building code calls for each room's required windows to have a total area of at least 10% of the floor area of the room or at least 10 square feet in area. Bathroom windows are required to have window area of at least 3 square feet. These must be manufactured with half of the window openable. Bathrooms without windows must be equipped with mechanical ventilation directly ducted to the outside.

Residential R3 building classification also requires that every dwelling have at least one bathroom containing a water closet, lavatory, and either a shower or bathtub. This room must also have separation from food preparation areas or storage rooms by a tight-fitting door. The water closet compartments must be at least 30 inches wide and have an unobstructed space of at least 24 inches in front of the water closet. Kitchens must contain at least one sink. All these fixtures (with the exception of the water closet) must be serviced with hot and cold domestic water from the dwelling's plumbing service side.

Attached garages must have 1-hour fire-resistive separation wall construction between garage and dwelling. Passage doors from dwelling to garage must be solid core and equipped with self-closing closure mechanisms as additional fire-resistive separation.

Minimum R3 residential floor space required is that one room must have at least 150 square feet of area, and bedrooms must have at least 70 square feet. No room, with the exception of a kitchen, can be less than 7 feet at any dimension. Ceiling height for R3 is set at a minimum of 7' 6" for at least 50% of the total room area and no part of the ceiling can be less than 5 feet from finish floor elevation. Smoke detectors are required in living and sleeping areas. Check with local code regulations for number and placement. R3 buildings three or more stories must be equipped with an approved manual or automatic fire alarm system. Most jurisdictions are now requiring fire protection systems for all new residential construction, regardless of stories or building height.

In the R1 multiresidential classification, change orders are often generated from a point of contention because the building code requires that each room have an exterior door to the outside; therefore they have immediate egress access unlike hotels. However, the building code still classifies them together. In this classification, corridors serving 30 or more people must have 1-hour fire-resistive construction. Doors leading into this corridor must be exterior-grade solid core with no grilles or louvers unless protected by a fusible link. Corridors exceeding 20 feet in length must have an egress door and cannot be terminated in a deadend.

Fire Detection Plans Checking

An approved automatic fire detection system with smoke detectors must be installed in accordance with NFiPA 72 and NFiPA 72E throughout all resident housing areas, including sleeping areas and any contiguous day room, group activity space, or other common spaces for customary access of residents. The only exceptions to this are:

- Smoke detectors are not required in sleeping rooms with four or fewer occupants in buildings protected throughout with an approved automatic sprinkler system.
- Other arrangements and positioning of smoke detectors may be used to prevent damage or tampering or for other purposes provided the function of detecting any fire is fulfilled and the siting of detectors is such that the speed of detection will be equivalent to that provided by the spacing and arrangements described in NFiPA 72E. This may include the location of detectors in exhaust ducts from cells, behind grills, or in other locations. The equivalent performance of the design must be approved by the building department.

Fire Protection Plans Checking

The fire resistance of building components such as structural framing, interior and exterior walls, nonloadbearing partition walls, roofs, floors, exterior doors, and windows determines the type of construction. If any of these primary building components is reduced in type of construction, the entire building is automatically classed in a lower classification. If one of these elements were reduced, a Type I building with required 3-hour structural frame, 2-hour floors, and 4-hour exterior bearing walls would be lowered in classification to Type V due to the lack of full compliance with Type I construction requirements. (See Table 12-1.)

The fire sprinkler piping may be connected directly to a domestic water supply system, but it must have a capacity sufficient to provide 0.15 gpm per sq. ft (0.0001 $m^3/s/m^2$) of floor area through the entire enclosed area. An indicating shutoff valve must be installed in an accessible location between the sprinkler and the connection to the domestic water supply. Smoke barriers must be provided in multiresidential construction to divide every story used by residents for sleeping, or any other story having an occupancy load of 50 or more persons, into at least two compartments. The exceptions here are spaces having direct exit to one of the following fulfill the requirements for subdivision of such spaces (provided the locking arrangements of doors involved meet the requirements for doors at the compartment barrier for the use condition involved):

- A public way, or
- A building separated from the resident housing area by 2-hour fire resistance or 50 ft (15.2 m) of open space, or
- An enclosed area having a holding space 50 ft (15.2 m) from the housing area that provides 6 sq. ft (0.56 m^2) or more of refuge area per person (resident, staff, visitors, etc.) that may be present at the time of the fire.

Doors in smoke barriers must swing in the direction of egress. In those applications where egress may be in either direction, a pair of swinging doors must be provided. Center mullions are permitted. The minimum clear width of exit in the direction of exit travel must be 32 inches (813 mm). Doors in a smoke barrier may be of the power-operated horizontal sliding type, provided the force to slide the door to its fully open position does not exceed 50 lb (222 N) with a perpendicular force against the door of 50 lb (222 N). The door must be reasonably tight fitting.

Portable space heating devices are prohibited. Any heating device other than a central heating plant must be so designed and installed that combustible material will not be ignited by it or its appurtenances. If fuel-fired, such heating devices must be chimney or vent connected, take air for combustion directly from outside, and be designed and installed to provide for complete separation of the combustion system from the atmosphere of the occupied area. The heating system must have safety devices to immediately stop the flow of fuel and shut down the equipment in case of either excessive temperatures or ignition failure. Approved suspended unit heaters may be used in locations other than means of egress and sleeping areas provided such heaters are located high enough to be out of the reach of persons using the area and provided they are vent connected and equipped with the safety devices called for above.

The fire resistance of building components such as structural framing, interior and exterior walls, nonloadbearing partition walls, roofs, floors, exterior doors, and windows determines the type of construction. If any of these primary building components is reduced in type of construction, the entire building is automatically classed in a lower classification. If one of these elements were reduced, a Type I building with required 3-hour structural frame, 2-hour floors, and 4-hour exterior bearing walls would be lowered in classification to Type V due to the lack of full compliance with Type I construction requirements.

Types of construction are classified by two factors, noncombustible and fire-resistive. Noncombustible materials will not burn, but they will transmit heat in a fire and contribute to flame spread, which maintains combustion if the fire becomes hot enough. Noncombustible materials include steel, masonry, iron, and concrete; however, any material which is exposed to enough heat will melt and disintegrate structurally. Fire-resistive materials will burn when exposed to prolonged flame. The building codes define these two properties, as applied to a building construction material, as either one of the following:

- Material of which no part will ignite and burn when subjected to fire. Any material approved as conforming to this definition is considered noncombustible.
- Material having a structural base of noncombustible material as defined above, with a surfacing material not over 1/8 inch thick which has a flame-spread rating of 50 or less as approved by the Fire Code.

"Noncombustible" does not apply to surface finish materials. Material requiring to be noncombustible for reduced clearances to flues, heating appliances, or other sources of high temperature refers to material conforming to Item 1 above. No material is classed as noncombustible which is subject to increase in combustibility or flame-spread rating, beyond the limits established, through the effects of age, moisture, or other atmospheric conditions. It should be noted here that there is a difference between flame-spread rating and fire resistive. A flame-spread rating compares the time it takes flame to spread on the surface of the tested material against the time it takes for the same amount of flame to spread on untreated, dried red oak of uniform density. On the flame-spread rating scale, red oak is given a rating of 100, representing maximum flame-spread. At the other end of the scale is cement asbestos board, which is given a rating of 0, representing minimum flame-spread. These specifications are separate and distinct and will invariably lead to change orders if not scrutinized in detail and applied exactly.

Fire resistance is determined by an Underwriter's Laboratories (UL) test of the material. A constant flame is applied to one side of the material in a controlled laboratory test environment and the temperature change (Delta factor) is measured on the other side. If, in the test time allotment, the exposed side does not exceed 250^{0} F, the material meets specifications for a UL-listed flame resistance and receives an according UL flame-resistance rating. Interior and exterior wall coverings as well as door and window assemblies are all sample tested and rated for fire resistance in this manner.

Remodeling

In addition to design work, there are three conventional ways to increase allowable floor area. The first is to install an automatic sprinkler system throughout which will increase allowable area in certain occupancies. The second way is to provide an increase in perimeter separation between the building and others adjacent to it on surrounding properties. And another is to build the structure using a higher type of construction. All of these methods incur more cost in per-square-foot construction or in extra land purchase, so in computing for allowable area increases, a balance must be struck between occupancy load and type of construction.

In determining allowable building height, remember that a roof structure designed to house anything other than mechanical service equipment will be counted as another additional building story. The height and area for buildings is governed by the intended occupancy or use of the building. Where a one-story automobile parking garage, enclosed or open of Type I or II construction, or open of Type III construction, with grade entrance, is provided under a building of Group R Occupancy, the number of stories to be used in determining the minimum type of construction may be measured from the floor above such parking area. The floor/ceiling assembly between the parking garage and the Group R Occupancy above shall comply with the type of construction required for the parking garage and shall also provide a fire-resistance rating not less than the occupancy separation required by code.

Where streets or public spaces, or horizontal separation from property lines of total width of not less than 30 ft (9144 mm), or 30 ft (9144 mm) between buildings on commonly owned property, extend along the building perimeter, except for hazardous occupancies, the areas permitted by code may typically be increased as follows:

$$I = 4/3[100\,(F/P - 0.25)]$$

Where

I = Percent increase of unsprinklered areas in Table 500
F = Building perimeter which fronts on streets, public

spaces, or horizontal separation not less than 30 ft (9144 mm) wide

P = Total perimeter of building

For both an unsprinklered building and a sprinklered building, the percent increase is multiplied by the unsprinklered area permitted for the type of construction of the building, and the resulting area increase is added to either the sprinklered or unsprinklered areas. When there are no unsprinklered areas permitted for the building, an unsprinklered area can be computed for use in this section. The corresponding unsprinklered areas are computed as one-third of the sprinklered area for one story only, and as one-half of the sprinklered area for multistories.

Where the exterior walls of two or more buildings located on the same lot face one another, and one of the walls is not constructed as required for a fire wall, a property line is assumed between them. The fire-resistance requirements for such facing walls and for the protection of openings therein must be the same as required by code for walls and openings facing an assumed property line. Fire-resistance separation normally is not required between a dwelling and its detached private garage.

Multiresidential Classifications

Because so much of the building code's rules involve the words "dwelling" and "residential," there have been many claims contentions in applying code regulations to the various types of "dwellings"—one-family houses, two-family houses, apartment houses, condominium units, dormitories, hotels, youth hostels, motels, etc. The present editions of the codes include terminology to eliminate such problems and use definitions of "dwelling" coordinated with the words used in specific code rules.

A "dwelling unit" is now defined as "one or more rooms" used "as a housekeeping unit" and must contain space or areas specifically dedicated to "eating, living, and sleeping" and must have "permanent provisions for cooking and sanitation." A one-family house is a "dwelling unit." So is an apartment in an apartment house or a condominium unit. But a guest room in a hotel or motel or a dormitory room or unit is not a "dwelling unit" if it does not contain permanent provisions for cooking, which must mean a built-in range or counter-mounted unit (with or without an oven).

Types of multiresidential construction are classified from the highest fire-resistive to the least, Type I being the most fire-resistive construction and Type V being the least. The type of construction will also determine the allowable building height. Building height limitations are given as maximum height in stories. It is important to note that due to high hazard situations, some occupancies are not allowed in specific types of construction regardless of area separation, floor area, or building height.

The classification of construction type is by the materials and fire resistance of its elements and/or the use of more than one construction type in a building. Every building is classified by the building department into one of the types of construction shown in Table 12-1.

Type I	**Type V**
Type II	1-Hour Protected
Type III	Unprotected
Type IV	**Type VI**
1-Hour Protected	1-Hour Protected
Unprotected	Unprotected

Table 12-1
Construction Type Classification

Materials Specifications Checking

In checking out the materials listed, keep in mind that all construction materials must bear identification showing the fire performance rating of that material. Such identification is issued and attached or marked on the material by the approved agency having a service for inspection of materials at the factory. Anything less is not "of equal or like kind."

Allowable openings in fire-rated walls are permitted only for small openings for ceiling conduit, piping, duct work, and electrical boxes made of ferrous metal. Maximum openings up to 100 square inches within 100 square feet of ceiling are permitted. However, fire-resistive floors must be continuous. Floor and roof penetrations as openings for electrical and mechanical equipment must be enclosed. The exception here is that vent and service piping may be installed through fire-resistive floors as long as the assembly does not reduce the required fire resistance rating of the floor. Conduit and pipes cannot be embedded in structural members or required fire protective coverings.

Protective coverings for penetrations in fire-rated assemblies must meet certain thickness requirements as set forth in the building code. These thickness requirements apply to all materials used from fire-resistive paint to sheetrock to concrete. These minimum thickness are measured in the material's net thickness and do not include any open or hollow spaces behind the protection. Know what these specs are. All fire-resistance construction requirements are expressed in terms of the number of hours of satisfactory performance in accordance with ASTM E 119. Construction required to have a fire-resistance rating must be supported by construction of equal or greater fire resistance. Type I construction is the most fire-resistant construction classification. Type I buildings are required to have 4-hour fire-resistive exterior walls. Most Type I construction nonbearing walls are allowed to have a 1-hour fire rating where unprotected openings are permitted. Nonbearing walls that front streets 50 or more feet wide can be of noncombustible, unprotected construction. Structural frames of Type I must be made of reinforced concrete, reinforced masonry, steel, or iron and have a 3-hour fire rating. Any load-bearing members that are components of an exterior wall must have the same fire-protection rating as the wall or meet a 4-hour fire rating. Type I is construction in which the structural members including exterior walls, interior bearing walls, columns, beams, girders, trusses, arches, floors, and roofs are of noncombustible materials and are protected to have fire resistance not less than that specified for the structural elements as specified by code.

Type II is construction in which the structural members including exterior walls, interior bearing walls, columns, beams, girders, trusses, arches, floors, and roofs are of noncombustible materials and are also protected to have fire resistance not less than that specified for the structural elements as specified by code. Type II construction requires all construction be of noncombustible components: concrete, iron, steel, or masonry. Nonbearing partition walls that are permanent may be constructed of fire-retardant-treated wood. A Type II 1-hour rating requires that all nonstructural components meet a one-hour fire rating. Type II-N has no fire ratings on any components, with the exception of the structural frame. In Type II construction, maximum allowable area and height is factored by the building's proposed occupancy rating.

Type III is construction in which fire resistance is attained by the sizes of heavy timber members (sawn or glued laminated) or by providing fire resistance not less than one hour where materials other than wood of heavy timber sizes are used; by the avoidance of concealed spaces under floors and roofs; by the use of approved fastenings, construction details, and adhesives for structural members; and by providing the required degree of fire resistance in exterior and interior walls. Heavy timber construction has certain requirements to fall under this classification. Roofs must be a minimum of 2 inches nominal thickness. Framed timber trusses and load-bearing columns must be nominal 8 inches by 8 inches to classify as heavy timber construction. Glue-laminated support beams that serve as floor supports must be a minimum nominal thickness of 10 inches by 8 inches, and glue-lam roof supports must be a minimum nominal thickness of 6 inches by 8 inches.

Type III construction requires that the building have 4-hour fire-rated exterior walls. In most occupancy groups, Type III construction limits the number of stories to two. Internal components, including structural frame, must meet only a 1-hour fire rating. In the 1-hour rated category, building height is limited to 65 feet, and Type III-N is limited to 55 feet in height. Wood columns may be sawn or glued laminated and cannot be less than 8 inches nominal in any dimension when supporting floor loads, and not less than 6 inches nominal wide and 8 inches nominal deep when supporting roof and ceiling loads only. Columns must be continuous or superimposed throughout all stories by means of reinforced concrete or metal caps with brackets; or be connected by properly designed steel or iron caps, with pintles and base plates; or by timber splice plates affixed to the columns by means of metal connectors housed within the contact faces, or by other approved methods.

Beams and girders of wood may be sawn or glued laminated and be not less than 6 inches nominal wide and not less than 10 inches nominal deep. Framed or glued laminated arches which spring from the floor line and support floor loads cannot be not less than 8 inches nominal in any dimension. Framed timber trusses supporting floor loads must have members of not less than 8 inches nominal in any dimension. Framed or glued laminated arches for roof construction which spring from the floor line and do not support floor loads must have members not less than 6 inches nominal wide and 8 inches nominal deep for the lower half of the height and not less than 6 inches nominal in any dimension for the upper half of the height. Framed or glued laminated arches for roof construction which spring from the top of walls or wall abutments, framed timber trusses and other roof framing which does not support floor loads, must have members not less than 4 inches nominal wide and not less than 6 inches nominal deep. Spaced members may be composed of two or more pieces not less than 3 inches nominal thick when blocked solidly throughout their intervening spaces or when such spaces are tightly closed by a continuous wood cover plate of not less than 2 inches nominal thickness, secured to the underside of the members. Splice places cannot less than 3 inches nominal thick. When protected by approved automatic sprinklers under the roof deck, framing members cannot be less than 3 inches nominal wide.

Wall plate boxes of this self-releasing type, or approved hangers, must be provided where beams and girders enter masonry. An air space of 1/2 inch (12.7 mm) must be provided at the top, ends, and sides of the member unless approved durable or treated wood is used. Girders and beams must be closely fitted around columns, and adjoining ends have to be cross-tied to each other (or inter tied by caps or ties) to transfer horizontal loads across the joint. Wood bolsters may be placed on top of columns which support roof loads only. Where intermediate beams are used to support floors, they must rest on top of the girders or be supported by ledgers or blocks securely fastened to the sides of the girders, or they may be supported by approved metal hangers into which the ends of the beams must be fitted closely. Columns, beams, girders, arches, and trusses of material other than wood must have a fire-resistance rating of not less than one hour. Wood beams and girders supported by walls required to have a fire-resistance rating of two hours or more must have not less than 4 inches (102 mm) of solid masonry between their ends and the outside face of the wall, and between adjacent beams.

Floors cannot have concealed spaces and must be constructed of sawn or glued laminated plank, splined, or tongue-and-grooved, not less than 3 inches nominal thick, or of planks not less than 4 inches nominal wide set on edge and well-spiked together. The planks must be laid so that no continuous line of joints will occur except at points of support and they cannot be spiked to supporting girders. Floor planking has to be covered with 1-inch nominal tongue-and-groove flooring laid crosswise or diagonally or with 15/32-inch (11.9-mm) wood structural panels. Planks and flooring cannot extend closer than 1/2 inch (12.7 mm) to walls to provide an expansion joint.

Roofs also must be constructed without concealed spaces and roof decks must be sawn or glued laminated, splined, or tongue-and-grooved plank, not less than 2 inches nominal thick, or of planks not less than 3 inches nominal wide, set on edge and spiked together as required for floors, or of 1-1/8-inch (29-mm) tongue-and-grooved wood structural panels bonded with exterior glue. Other types of decking may be used when approved by the local building department. Bearing portions of exterior and interior walls must be of approved noncombustible materials and provide the fire-resistance ratings in accordance with code requirements. Nonbearing portions of exterior walls also must be of approved noncombustible materials and provide the fire-resistance ratings in accordance with code requirements. Where a horizontal separation of at least 20 ft (6096 mm) is provided, wood columns, arches, beams, and roof decks conforming to heavy timber sizes may be used externally.

Type IV is construction in which the structural members including exterior walls, interior bearing walls, columns, beams, girders, trusses, arches, floors, and roofs are of noncombustible materials. Type IV construction may be protected or unprotected. Type IV construction requires that wood floors be ventilated in the underfloor areas, as this type of construction is usually not built on slabs as the other types but rather on perimeter foundations with pier and posts supporting wooden girders.

Type V is construction in which the exterior bearing and nonbearing walls are of noncombustible material. Type V construction allows a wood frame if the exterior walls are not required to be noncombustible. Type V one-hour requires a one-hour fire rating for all construction components. Most R3 and multifamily residence buildings are required to be of Type V-N construction.

Type VI is construction in which the exterior bearing and nonbearing walls and partitions, beams, girders, trusses, arches, floors, and roofs and their supports are wholly or partly of wood or other approved materials. Type VI construction may be either protected or unprotected. Fire-resistance requirements for the structural elements of Type VI construction must be as specified in the adopted local building code.

Interior Finishes Plans Checking

Although the code consists essentially of specific regulations on details of interior design and installation, there is occasional explanatory material in the form of notes to rules that must be understood in checking plans for potential interior change orders. Compliance with the code consists in satisfying all requirements and conditions that are stated by the use of the word "shall." That word, anywhere in the code, designates a mandatory rule. Failure to comply with any mandatory code interior finish rule constitutes a code violation and will lead to a change order before the inspection will pass.

In finish work, most contracts call for craftsmanship of a standard "workmanlike manner." This statement has been the source of many claims because opinions differ as to what is an installation according to a "workmanlike manner." The building code places the responsibility for determining what is acceptable and how it is applied in the particular jurisdiction on the authority having jurisdiction. This basis in most areas is the result of:

- Competent knowledge and experience of installation methods
- What has been the established practice by the qualified journeyman in the particular local area
- What has been taught in the trade in the trade schools having certified training courses for apprentices and journeymen

Codes and standards must be carefully interrelated and plans checking here must be followed with care and precision. Modern finish work that fulfills these demands should be the objective of all of builders and designers. The steps in plans checking for interior design construction to code compliance are:

1. Determine the occupancy classification for the building's design purposes.
2. Determine the fire-resistive hour rating needed for the occupancy of the building.
3. Determine the type of construction that meets the required fire rating.
4. Determine the materials that meet compliance with the type of construction required.

Combustible materials may be used as a finish for ceilings, floors, and other interior surfaces of buildings as provided by code. Windows in the first story of buildings may be of wood or of unprotected metal framing. Interior finish includes the exposed interior surfaces of buildings, including, but not limited to, fixed or movable walls and partitions, columns, and ceilings, interior wainscoting, paneling, or other finish applied structurally or for decoration, acoustical correction, surface insulation, structural fire resistance or similar purposes. Requirements for interior finishes do not apply to trim (defined as picture molds, chair rails, baseboards, and handrails), to doors and windows or their frames; nor to materials which are less than 1/28 inch (0.9 mm) thick cemented to the surface of walls or ceilings, when these materials have flamespread characteristics no greater than paper of this thickness cemented to a noncombustible or fire-retardant treated wood backing. Interior finish materials other than those applied to floors are classified in accordance with ASTM E 84. Such interior finish materials are grouped in the following classes in accordance with their flamespread and smoke development:

- Class A Interior Finish. Flamespread 0-25, Smoke Developed 0-450. Any element thereof when tested does not continue to propagate fire.
- Class B Interior Finish. Flamespread 26-75, Smoke Developed 0-450.
- Class C Interior Finish. Flamespread 76-200, Smoke Developed 0-450.

In buildings of Type I or Type II construction, any floor finish made of combustible material must be applied directly upon the floor construction, except that a floor finish of wood, linoleum, rubber, tile, or cork may be secured to a subfloor of wood. Certain extra work orders are overlooked here. Where wood sleepers are used for laying wood floors or subfloors in such buildings, they must be fireblocked so that there will not be an open space extending under any permanent partition. Where wood sleepers are used and the space between the floor slab and the underside of the floor or subfloor is more than 2-1/2 inches (64 mm), that space must be filled with noncombustible material so that such space is not more than 2-1/2 inches (64 mm). Combustible insulating boards may be used for sound deadening or insulating of floors, except that in buildings required to be of Type I or Type II construction, such insulating board cannot be more than 1/2 inch (12.7 mm) thick and cemented directly to the floor slab or secured to wood sleepers fireblocked as called for above and covered with approved finish flooring.

Finished floors or floor covering materials of a traditional type—such as wood, vinyl, linoleum, terrazzo, and other resilient floor covering materials—are exempt from these requirements. Carpet floor coverings are tested as proposed for use including underlayment. Carpet materials used on floors of exit access corridors and enclosed exits in other than Group I Occupancies must satisfactorily withstand an ignition factor known as a minimum critical radiant flux of 0.22 watts/sq. cm when tested in accordance with the NFiPA 253. The exception to this is a building equipped with an approved automatic sprinkler system. The carpet must be identified by a hang tag or other suitable method as to identify manufacturer or supplier and must indicate the critical radiant flux level. Be sure the numbers match the specifications. When walls and ceilings are required by code to be of fire-resistant construction, noncombustible or fire-retardant-treated-wood construction, the finish material can be applied to furring strips not exceeding 1-3/4-inch (44 mm) thick applied directly against such surfaces. The intervening spaces between such furring strips must be filled with inorganic or Class A materials or fireblocked not to exceed 8 ft (2438 mm) in any direction. Where walls and ceilings are required to be of fire-resistant construction, noncombustible or fire-retardant-treated-wood construction and walls are set out or ceilings are dropped distances greater than specified in code, Class A finish materials must be used except where the finish materials are protected on both sides by automatic fire-extinguishing systems or are attached to a noncombustible or fire-retardant-treated wood backing or to furring strips installed. The hangers and assembly members of such dropped ceilings that are below the main ceiling line must also be of noncombustible or fire-retardant-treated-wood materials.

Wall and ceiling finish materials of all Class A, B, or C materials may be installed directly against the wood decking or planking of heavy timber construction or to wood furring strips applied directly to the wood decking or planking installed and firestopped. Interior finish materials must be cemented or otherwise fastened in place so that they will not readily become detached when subjected to room temperatures of 300^0 F (149^0 C) for 25 minutes.

Project Duration

Straight-line time-scaled bar charts and CPM management are the best way to plot out and execute a residential. Both systems touch on all phases of the construction and components installation, but CPM is a more effective means of totally viewing a project for potential change orders when compared to bar charts. However, if the estimate was done poorly, the estimate was made at the job logic level by estimating the labor hours needed to complete each work item, but no consideration was taken that actual field construction has practical constraints, and no real-world cushioning allowance was included. So float in the CPM project schedule has to be used efficiently to bring all the labor, materials, and fabrications together at the proper time. But the original estimate should show no float. You can best control your estimates by comparing them with others and, if prices vary significantly, by verifying the numbers.

After running the first pass on the computer, it's a good idea to recheck the elapsed times of the critical and near-critical activities to ensure that the time estimates are reasonable for components delivery and installation. The key critical activities are the 20% or so that most directly affect the project completion date. You should also consider any abnormal factors, such as conflicting work schedules, adverse weather conditions, shortages of material and human resources, and delivery dates that could adversely affect an otherwise normal elapsed-time estimate. The builder acting as prime contractor is typically not allowed delay claims for unseasonable or inclement weather. Keep in mind, however, that field labor productivity reflects directly on elapsed-time estimates. As an example, the basic activities (with their average elapsed times in estimated working days) to site-build a typical residential house in the Midwestern states are listed in Table 12-2.

Activity	Elapsed time
Clear site	10
Excavate basement	15
Rough walls and floors	45
Exterior siding	35
Pour basement	30
Roof	20
Rough-in plumbing	25
Rough-in electrical	20
Finish plumbing	20
Finish electrical	14
Interior painting	30
Drywall, taped and textured	27
Exterior fixtures	10
Exterior paint	30
Finish flooring	25
Roofing	20
Landscaping	30

Table 12-2

Build your own sample project schedule for these activities by arranging the work activities according to their job logic requirements. Now draw an arrow diagram from each activity, showing connectivity and sequential flow. Next calculate the early and late start dates, with the project starting any time you wish. Then calculate the earliest completion time and the critical path. Finally, interdependencies and constraints must be identified between activities, with float used to balance constraints. This is the basic format in elapsed-time estimating in CPM project scheduling. Granted, this is a simplistic example, but regardless of the size of the project the fundamentals remain the same. With bigger projects, you'll just have a lot more activities.

If phased construction options were used, the duration could be shortened to look like Table 12-3.

Activity	Elapsed time
Clear site	10
Excavate basement	15
Rough walls and floors	15
Exterior siding	0
Pour basement	30
Roof	1
Rough-in plumbing	0
Rough-in electrical	0
Finish plumbing	10
Finish electrical	7
Interior painting	0
Drywall, taped and textured	3
Exterior fixtures	3
Exterior paint	0
Finish flooring	0
Roofing	0
Landscaping	30

Table 12-3

By line-item comparison, you can see the activities completion times are shortened as follows in Table 12-4.

Activity	Standard	Phased	Days saved
Rough walls & floors	45	15	30
Exterior siding	35	0	35
Exterior siding	35	0	35
Roof	20	1	19
Rough-in plumbing	25	0	25
Rough-in electrical	20	0	20
Finish plumbing	20	10	10
Finish electrical	14	7	7
Interior painting	30	0	30
Drywall, taped and textured	27	3	24
Exterior fixtures	10	3	7
Exterior paint	30	10	20
Finish flooring	25	0	25
Roofing	20	0	20

Table 12-4

Even if the activity events in the site-built example are dovetailed properly with parallel activities done sequentially, the completion time saved in working days by using modular construction components is substantial. To produce logical change order situations, make a first draft of your production schedule by simply using the schedule outline in Figures 12-1 through 12-3 by using related activities to your job and fill in with your calendar dates. All production schedules, no matter how complex, begin like this by listing the activities and scheduling the dates. These are standard activities, in typical job logic, for an average SFD home. You may copy this directly and supply your project's dates to begin a specific project schedule.

Item Date completed	Schedule date	Start date
Flag lot		
Clear and grub		
Stake out		
Excavate		
Form footing		
Inspection: Footing		
Pour footing		
Order lumber drops		
Wall forms/panels		
Inspection: Walls		
Pour walls		
Groundworks		
Order waterproofing		
Inspection: Groundworks		
Order soil treatment		
Prep basement slab		
Drain tiles		
Inspection: Slab		
Pour basement slab		
Frame basement		
Inspection: Exterior drain tiles		
Backfill/rough grade		
Form areaway steps/pad		
Inspection: Areaway		
Pour areaway		
Order mechanical fabrication		
Structural framing		
Set roof trusses		
Roof sheathing		
Roof covering		
Install insulating sheathing		
Set windows		

Figure 12-1

Set exterior doors
Insulate around doors/windows
Set meter base
Schedule power
Schedule gas service
Window/door trim
Install stone work
Rough in HVAC
Rough in electrical
Rough in plumbing
Check framing
Carpentry punch out
Exterior siding
Install prefab fireplace(s)
Order electric meter
Order gas meter
Install furnace
Rough clean
Inspection: Framing
Inspection: Rough plumbing
Inspection: Rough electrical
Inspection: HVAC
Airstop/firestop
Insulation
Order drywall stock
Inspection: Close-in
Confirm cabinets
Order interior trim
Hang drywall/finish
Schedule water meter
Install interior trim
Form exterior concrete
Pour exterior concrete
Fine grade
Gutter/downspouts
Wrought iron rails

Figure 12-2

Interior paint
Install cabinets
Vinyl floors
Hardware floors
Ceramic tile
Install appliances
Final plumbing
Final electrical
Start-up HVAC
Sod/landscaping
Grading/lot certification
Door hardware
Splash blocks
Point-up drywall
Touch-up paint
Final clean 1
Carpet
Install screens
Final clean 2
Inspection: Final

Figure 12-3

Sample Forms

These forms are included for you benefit and can be used as templates for your own documents. Figure 12-4 is a sample claims checklist that could be used on the previously exampled SFD project. Figure 12-5 is a generic change order, standard AIA Doc G701. Figure 12-6 is a detailed change order that can be customized with your company's letterhead.

SAMPLE CONSTRUCTION CLAIMS CHECKLIST

People necessary for information and/or testimony

Key project people:____________________

Subcontractors involved:____________________

Project owner:____________________

Project manager:____________________

Project engineer:____________________

Project superintendent:____________________

Scheduler:____________________

Architect:____________________

Engineers:____________________

Lawyers/experts/consultants:____________________

Accountants:____________________

Other:____________________

Key documents (originals)

Plans, specifications, addenda:____________________

Related files:____________________

Notice date establishment:____________________

Pass-through reference:____________________

Change clause:____________________

Dispute clause:____________________

Arbitration clause (agreement to arbitrate):____________________

Other:____________________

Project records

Contract/subcontract/P.0:____________________

Change order files:____________________

Change order summary log:____________________

Daily field reports - From:__________ To:__________

Schedule: Baseline (target):____________________

Updates:____________________

Projections/analysis:____________________

Narratives:____________________

Job cost records:____________________

Shop drawings and transmittals:____________________

Shop drawing log:____________________

Other:____________________

Figure 12-4
Sample Claims Checklist

CHANGE ORDER

AIA DOCUMENT G701

Distribution to:

OWNER	[ ]
ARCHITECT	[ ]
CONTRACTOR	[ ]
FIELD	[ ]
OTHER	[ ]

PROJECT: CHANGE ORDER
NUMBER:
(name, address)
INITIATION DATE:
TO (Contractor):
ARCHITECT'S PROJECT NO:
CONTRACT FOR: CONTRACT DATE:
You are directed to make the following changes in this Contract:

Not valid until signed by both the Owner and Architect. Signature of the Contractor indicates his agreement herewith, including any adjustment in the Contract Sum or Contract Time.
The original (Contract Sum) (Guaranteed Maximum Cost) was..................................$
Net change by previously authorized Change Orders..$
The (Contract Sum) (Guaranteed Maximum Cost) prior to this Change Order was...........$
The (Contract Sum) (Guaranteed Maximum Cost) will be (increased) (decreased) by ....$
The new (Contract Sum) (Guaranteed Maximum Cost) including this change will be........$
The Contract Time will be (increased) (decreased) (unchanged) by......................... _________ Days.
The Date of Substantial Completion as of the date of this Change Order therefore is ______________
Authorized:

ARCHITECT	CONTRACTOR	OWNER
Address	Address	Address
BY ______________	BY ______________	BY ______________
DATE ______________	DATE ______________	DATE ______________

Figure 12-5
Generic Change Order

LETTERHEAD

Date

To: *Owner or architect, per specification direction*

Re: *Project number*
Job name/description
Change No:
Or subcontract amendment #

Gentlemen:

Your proposal dated ____________ in the amount of $ __________________
for ____
____
____
____
has been accepted and incorporated into contract no. ________ dated _______
as Change Order Number ________________________ . The contract is being (increased / decreased) by the sum of $ ________________________ .
All other conditions of the contract remain unchanged.

1. Original Contract Amount $
2. Previous Additions $
3. Previous Deductions $
4. Previous Total $
5. This Change Order $
6. New Contract Amount $

The Contract Time will (remain unchanged / be increased / be decreased) by _______ (working / calendar) days, and is to be complete on _____________.

ACCEPTED BY: ____
Date ________________________________

Figure 12-6
Detailed Change Order

Chapter 13

Advanced Claims Management

Claims and Disputes

The legal term *claim* applies to differences that are developed during the life of the contract under protests and potential claims, and that are not yet resolved at the time the contractor returns the proposed final estimate of the amount of additional money or time asked for. In other words, a protest or potential claim does not become a claim until the contractor repeats its objections by notifying the architect/engineer or owner at the time the proposed final estimate for such claim is returned to the architect/engineer.

Whether it is the competitive bidding process, increased competition, or just part of the growing trend toward more litigation, more and more projects are being affected by claims and disputes. Quite often, a majority of contractor claims and disputes arise out of poorly drafted or ambiguous contract documents.

Disagreements between the contractor and the owner, and increasingly including the architect/engineer, regarding interpretation of the plans and specifications, what should be considered as extra work to the contract, payment for contract work and change orders, extensions of time, damages for delay caused by either the owner or the contractor, changed or unforeseen conditions, performance of subcontractors, compliance with contractual requirements, partial acceptance of a project by the owner, and similar problems seem to be increasing nationally on job sites of all sizes.

When such claims or disputes arise, the contractor, owner, and architect/engineer must pay careful attention to the procedures set forth in the contract documents for the legal handling of construction claims. Claims management in your contractual obligations, therefore, is a critical factor in ensuring your business longevity.

Owner Notification

Typically, contract documents require a contractor to alert the owner and/or the architect/engineer immediately, or within a certain number of days, of any potential claim or dispute when it arises. Contracts normally require that any claim or dispute must be submitted to the architect/engineer or owner's representative as the first step in the claims resolution process. Contracts often set forth an appeals procedure so that the contractor can proceed with its claim or dispute if the architect/engineer's decision is unfavorable. When all administrative steps set forth in the contract documents have been exhausted, the contractor may then, and only then, resort to the courts or arbitration, if provided for under the contract. The procedural steps set forth in the contract must be followed by the contractor. Failure to follow these steps could in itself be considered a breach of the contract by the contractor.

Administrative Procedures

A contractor must exhaust all of its administrative remedies before resorting to the courts and must strictly adhere to any claim procedures mandated by law. For example, in addition to the claims and disputes provisions outlined in the typical owner contract, some states require that a claim first be filed by the contractor against the appropriate owner, with that owner subsequently rejecting the claim within a prescribed time, before the contractor has the right to take the matter to court. A contractor's failure to follow mandated government claims procedures can result in the contractor being barred from recovery on its claim.

Work Performed under Protest

Except for total nonpayment by the owner, or due to particular circumstances under a specific contract, the contractor cannot refuse to proceed with the work on the project when a claim or dispute arises. Normally, the contractor must continue its operations despite any claims or disputes. Generally, the contractor must perform any disputed work under protest and must still complete the project on schedule, leaving any unresolved claims or disputes until after the project has been completed.

When a claim or dispute arises, or when the contractor foresees that it will arise, the contractor must take two immediate steps. First, the contractor must give the owner notice of any claim or dispute, or any potential claim or dispute, as soon as possible. Most contracts contain very specific requirements for "Notice of Claims" or "Notice of Potential Claims" that set forth specific time frames within which a contractor must notify the owner. Although the courts have varied in their interpretation and enforcement of such time requirements, a contractor runs a risk when its claims are not filed in accordance with contractual requirements.

Owner Opportunity to Correct

Regardless of any contractual language regarding the time for making claims, common sense and equitable principles dictate that once a contractor becomes aware of a claim or a potential claim, it must be communicated to the owner. For example, if the owner's actions in some way delay the contractor and the project, the owner must be advised of the situation in order for the owner to have an opportunity to correct or alleviate the problem. If a contractor does not give notice to the owner that it is being delayed at the time that the delay is occurring, the contractor will have very little chance of recovering damages incurred as a result of such delay after the project is complete. Without proper notice, the owner will argue, and rightfully so, that it cannot be held liable to the contractor for a delay that it was never made aware of and never had an opportunity to do anything about.

Contractor's Right to File Claims

Contractors are entitled to file any real and provable claim that arises during the course of a project. Claims should not be made or taken "personally," as they are simply the means available to the parties to the contract to be able to adjust the contractual and economic relationship between them to meet changing conditions. However, the filing of claims is not a proper way for the contractor to make up for a bad bid, or to recoup losses incurred on the project. Ultimately, it is the integrity of the contractor that dictates what claims are and are not made against an owner. In general, if a contractor only presents claims to an owner that are legitimate and verifiable, it will stand a better chance of having those claims, and all other claims it files, settled in an equitable manner. If, on the other hand, the contractor makes it a practice to file claims for any and all reasons, some less legitimate than others, the contractor runs the risk of prejudicing its legitimate claims and reducing the possibility of resolving them without litigation's very costly and delayed process for both parties.

A contractor is entitled to pursue every legitimate claim that it has, but must recognize that the owner is entitled to proper notice and adequate proof of the claim before the owner has any obligation to pay on the claim. Contrary to a bidding situation where the contractor assumes only the best-case circumstances, a contractor's claim is generally estimated by assuming only the worst-case circumstances in the determination of estimated impact costs yet to be incurred. Particular attention should be paid to situations where claims are directly related to the actions or inactions of the owner's representatives or agents. Claims that sometimes arise as a result of the obligations of the owner's architect/engineer to design a portion of a project properly are assertable against the owner, and under some circumstances, directly against the architect/engineer. Although claims often provide the contractor with an opportunity to attempt to make up for mistakes made in its bid or in the construction of the project, it is not a good practice for a contractor to make more claims simply to make up for mistakes and errors. No purpose will be served by a contractor who turns a project into a battleground to cover losses incurred as a result of its own mistakes and errors.

Dispute Between Parties

In general, most claims issues will fall into approximately ten classifications, with numerous variations possible within each category. They are:

1. Owner-caused delays in the work.
2. Owner-ordered scheduling changes.
3. Constructive changes.
4. Differing site conditions.
5. Unusually severe weather conditions.
6. Acceleration of the work; loss of productivity.
7. Suspension of the work; termination.
8. Failure to agree on change order pricing.
9. Conflicts in plans and specifications.
10. Miscellaneous problems.

Owner-Caused Delays

The majority of all claims involve at least some elements of delay, even if the primary issue is one of the other categories. Most contractors fail to realize the potential for recovery of losses in this sensitive claims area, until they have consulted a claims specialist.

Delay claims fall into three categories: nonexcusable, excusable, or compensable. Nonexcusable delay is one that is caused by factors within the contractor's reasonable control. Essentially, this means that the delay is the contractor's fault, so the contractor will be unable to recover additional time or additional compensation. A delay that is caused by factors beyond the contractor's reasonable control, but is not the result of the owner's actions or failure to act, is considered excusable. An excusable delay entitles the contractor to an extension of time but to no additional compensation for the cost of the delay.

Contractor Entitled to Complete Early

An unusual twist on the delay issue occurs when a contractor plans to finish a project ahead of schedule. In some cases, an owner has been known to object to early completion of a project for varied reasons. However, it should be kept in mind that an owner may be held liable for hindering early completion of a project [Appeal of CWC, Inc., ASBCA No. 264342 (June 29, 1982)]. CWC, Inc., was prime contractor for renovation of certain military buildings. The contract required the government to arrange access to the individual buildings within two weeks after receiving a request from the contractor. In several instances, however, the government failed to provide access in a timely manner.

The contractor finished the entire project in less than the allowed time and prior to the completion date indicated in the contractor's own schedule. Nevertheless, the contractor submitted a claim for the increased costs caused by the government's delay. The Armed Services Board of Contract Appeals recognized the claim and issued the following statement: "Barring express restrictions in the contract to the contrary, a construction contractor has the right to proceed according to his [or her] own job capabilities at a better rate of progress than represented by his [or her] own schedule. The government may not hinder or prevent earlier completion without incurring liability." The bottom line here is that if the owner has reason to require the contractor to conform to a scheduled completion date, allowing neither delays or early completion, this requirement must be clearly spelled out in the contract documents.

Sources and Causes of Time-Related Disputes

All too often, owners and architect/engineers automatically blame the contractor for virtually all construction delays. Although that is often a well-founded accusation, there are many sources and causes of time-related disputes, as indicated in the following list:

Source Cause Owner
1. Lack of expertise
2. Long line of authority in project organization
3. Delayed approvals of schedules and change orders
4. Slow change order processing
5. Failure to obtain permits
6. Irrelevant milestone dates in documents

Contract Documents
1. Inadequate scheduling clauses
2. Directing work sequence by owner or architect/engineer
3. Drawings not indicating interfaces
4. Permitting responsibilities vague
5. Milestone dates and interface clauses
6. Leverage for enforcement of schedule specification
7. Coordination inadequately specified

Construction Manager
1. Lack of expertise in schedule management by designer
2. Implementation of specifications
3. Inadequate record keeping
4. Inadequate schedule updates and progress monitoring
5. On-site coordination
6. Job meetings

Contractor
1. Noncompliance with specifications

2. Schedule updates not done
3. Reluctance to cooperate and coordinate
4. Failure to meet milestone dates
5. Not following permit requirements

Compensable versus Noncompensable Delays

A delay that is considered compensable is one where the owner has failed to meet an obligation stated or implied in the construction contract. If a delay is considered compensable, the owner must grant a time extension and reimburse the contractor for the increased cost caused by the delay. If a contractor experiences concurrent delays where one is compensable and the other is merely excusable, no compensation is allowed. Similarly, on concurrent delays where one is caused by the contractor and the other by the owner, the delay is considered neither excusable nor cornpensable to the extent of the overlap or concurrency [Appeal of Rivera General Contracting, ASBCA No. 25888 (April 30, 1985)].

It should be stressed that these definitions are general in nature. The parties have great latitude to contractually determine whether or not the contractor will be entitled to additional compensation, an extension of time, or nothing at all. If an event delays an activity on the critical path, the entire project completion date will be delayed. To avoid delay, it is necessary to accelerate the performance of a subsequent critical activity or adjust the sequence of work activities to compensate for the delay. If a delay occurs to an activity not on the critical path, however, the impact is cushioned by the available free-float time. Delays to a noncritical activity, which do not exceed the amount of free-float for that activity, will not result in a delay to the completion date of a project. If, however, a delay to a noncritical activity directly caused an increase in the contractor's costs, such costs may be recovered by the contractor if the delay was the owner's responsibility. This, of course, is a matter requiring very clear proof to overcome the presumption that delays to noncritical activities do not result in delay-related costs. Although an owner-caused delay to a noncritical activity that used only free-float time may not be justification for an extension of time, such delays may still be compensable when they cause the contractor to accelerate the work, cause inefficiency or disruption, or delay some specific activity and thus cause extended costs, such as cost of equipment rental. In addition, such delay is not necessarily measured from the time that the contract documents required the project to be completed, but rather from the time that the contractor could have completed the project, from a completion date earlier than the one indicated in the contract documents [Metropolitan Paving Co. v. United States, 325 F.2d 241 (1963)].

The most valuable tools for the defense of a claim are properly prepared CPM schedules. Delays are the most prevalent problem on construction projects. A variety of factors in an ongoing project can lead to a delay. Typical of the types of delays that might be caused by an owner include the following:

1. Late approval of shop drawings and samples
2. Late approval of laboratory tests
3. Delays in answers to field inquiries by the contractor
4. Changes in the contractor's method of doing the work
5. Variation in estimated quantities
6. Interference with the contractor during construction
7. Owner-caused schedule changes
8. Design changes
9. Changes in inspection level
10. Failure to provide for site access
11. Lack of required rights-of-way
12. Interference by other contractor's or owner's forces

A contractor may jeopardize its claim, however, by being the cause of nonexcusable delays of its own. Such delays are frequently the result of any of the following causes:

1. Late submittal of shop drawings
2. Late procurement of materials or equipment
3. Insufficient personnel
4. Unqualified personnel
5. Inadequate coordination of subs or other contractors
6. Subcontractor delays
7. Late response to owner and architect/engineer inquiries
8. Construction not conforming to contract requirements, making repeated reworking necessary

Attendance at Site Visit

From time to time, owners and architect/engineers feel that they will better protect the owners interests by requiring all bidders to attend a prebid site visit. This is obviously based upon the presumption that the contractor might have a stronger claim for differing site conditions if he or she failed to attend a scheduled prebid site tour. However, it appears that the owner may not be as vulnerable as at first thought. In a 1986 case, the Comptroller General ruled that a contractor's failure to attend a prebid site visit is not a valid reason to reject an otherwise acceptable low bid [Matter of Arrowhead Construction, Inc., Comp. Gen. No. B-220386 (January 8, 1986)]. In the subject case, the government requested bids for construction of a paint booth. The Notice Inviting Bids required all bidders to attend a site visit on a designated date. When the government announced its intention to award the contractor to a bidder that did not attend the site visit, the second-lowest bidder protested. The Comptroller General denied the protest.

The decision was that a bidder that fails to attend a prebid site visit does so at its own risk and will be held responsible for any information that it could reasonably determine from that inspection. According to the Comptroller General, the site inspection does not relate to the responsiveness of the bid. Even when the Notice Inviting Bids expressly requires attendance at a prebid inspection, failure to attend is not a valid basis for rejection of an otherwise responsive bid.

Outdated Standards

Perfect specifications are hard to find. Actually, I only say that because authors tend to write in generalities. I, nor any of the other contractors I have worked with over the years, have never seen perfect specifications. So the contractor must make a reasonable interpretation at the time of bidding the job. A conscious effort here will put the contractor in a good position for monetary recovery. Frequently, the contractor will find in the specifications that outdated standards are specified. Another common oversight are products specified that no longer exist. Often, the specifications will contain references stating that wherever codes or commercial standards are specified, the contractor is obligated to use the latest issue of that standard existing at the time that the project went to bid. What commonly occurs in many instances is that the designers considered a design based upon an old standard that exists in their files, or a standard that was current during the design phase, but which later may have been updated by the sponsoring agency/manufacturer without the designer's knowledge. Occasionally, serious difficulties arise from such practices, and the contractor certainly should have the right to the project cost difference resulting from the error. The acceptance of shop drawings that deviate substantially from the requirements of the contract documents must be accompanied by a written change order.

Permits

A building permit is a license to proceed with the work but not authority to violate, cancel, alter, or set aside any of the provisions of the technical codes. Issuance of a building permit will not prevent the building department from thereafter requiring a correction of errors in plans, construction, or violations of this code. Claims for permit responsibilities are often generated from these additional corrections. The contractor must be aware that every building permit issued becomes invalid unless the work authorized by such permit is commenced within 6 months after its issuance, or if the work authorized by the permit is suspended or abandoned for a period of 6 months after the time the permit is issued.

Cost Projections

If cost projections are being prepared by manual means, one way to keep track of the change-order costs is to assign a separate code to each change order. Then all costs associated with the change order will receive the same cost-code number. For example, 09900 series cost code can be set aside for change orders. The first change order for a particular project can be assigned cost code 09901, and change order 2 will be assigned cost code 09902. All costs incurred to complete the work outlined in change order 1 will therefore be coded 09901 and will be posted against the estimated cost of the work.

This system may not work when electronic data processing is being used. Problems may arise when a change order to an owner contains work that will be subcontracted to several different trades. A change order involving work by multiple trades must be cost-coded so that each trade's contract is updated.

Let's say that a change order requested by an owner amounts to $10,000 without the contractor's overhead and profit, and involves HVAC work worth $6500, plumbing work for $1000, and electrical work amounting to $2500. When a change order to the owner is issued for $10,000 plus overhead and profit, the cost of this work must be distributed where it belongs, and all trades receiving a change order for this work will have that portion of the change relating to their work cost-coded to their original contract. If the HVAC subcontractor's contract has been coded 15500, it is possible to use the last two digits to highlight changes to that contract; therefore, change order 1 to his account would be coded 15501.

General Conditions of the Contract

The industry prevailing standard construction contract is the American Institute of Architects (AIA) General Conditions contract (Document A201). In this section we will examine the articles relative to the claims management process, Articles 6 through 9. The actual change order article is Article 7: Changes in the Work, which deals with the process in depth. A201 states that a change order is defined as a written order to the contractor, having been signed by the owner and architect, issued after the contract has been signed and authorizing a change in the scope of the work or an adjustment in the contract sum or contract time, or both. Because the general contractor's goal will be to complete a project on time, or submit a change order to request an extension in contract time, if the contract contains a liquidated-damages clause, these two seemingly unconnected aspects of the construction process do possess a common thread. One major problem in change order work concerns the disparity between the theoretical way change orders should be processed and the practical manner in which they are usually handled. Article 7 of the AIA General Conditions document states that a change order is a "written instrument prepared by the architect and signed by the Owner, Contractor, and Architect stating their agreement" to a change in the work, the cost of this work, and the extension of time, if any, required to put this work in place. If one follows the rules, a change order's life starts when the architect submits a proposal request in which it is requested that the contractor submit a proposal for all costs to perform the work outlined in the request. Each such proposal is numbered sequentially by the architect, and this number is referred to in the contractor's cost proposal or estimate. After this cost estimate is reviewed and approved by the architect, a formal change order, usually AIA Document G701, is prepared and sent to the contractor for signing. It is then returned to the architect, who signs it and sends it on to the owner for signature. When this fully executed document is returned to the contractor, only then is the contractor is authorized to proceed with the work. This is a rather lengthy and time-consuming process by which to get the work done, contrasting the normal urgency that change order work has about it.

This provision takes on added importance if the contractor agrees to certain terms that are not industry standards. For instance, a one-year warranty is standard in the industry, but if the owner has negotiated a two-year guarantee on mechanical equipment and the general contractor has not negotiated a similar guarantee from the mechanical subcontractor, the general contractor will be responsible for that two-year warranty. Article 5.3.1 goes even further. On the request of a subcontractor, the general contractor is to identify the terms and conditions of the proposed subcontract agreement and how these terms and conditions vary from the contract between owner and general contractor. A new requirement in the A201 document requires the general contractor to insert an "assignment" clause in each subcontract agreement. In case of the termination of contract between owner and builder, these subcontracts would be assigned, or taken over, by the owner.

Article 6 relates to construction by owner or by separate contracts. The owner reserves the right to perform certain construction work with his own forces or other subcontractors. Even though the owner has a right to perform work in his building under construction, this right should not go unchallenged. A union contractor may experience problems if the owner selects a nonunion subcontractor to work alongside union workers. Although job actions may not be legally defensible, they could create temporary work stoppages or slowdowns and have a serious effect on a tight schedule.

Coordination is another concern, and although the owner is to provide coordination for work forces he employs, if he is not a sophisticated purchaser of construction work, other problems can, and *will*, arise. For instance, who will cut and patch for these owner's subcontractors? Will they severely damage finished surfaces and then state that their contract with the owner precludes repair? The contractor should attempt to force cooperation by the owner in reviewing the scope of work being considered prior to the owner's award of contract in such cases. Expert advice at this time will make for good public relations and lessen the chances of litigation later on.

The owner's right to keep the premises clean permits awarding a contract for cleaning per paragraph 6.3.1 of this article. If a dispute arises with the general contractor, these costs will be backcharged to the contractor in an amount determined by the architect.

Article 7 covers changes in the work. A new term was introduced into the 1997 version of Article 7, construction change directive, which is defined as "an agreement by the owner and architect that may or may not be agreed to by the contractor." This construction change directive is a written order prepared by the architect and signed by the owner directing the contractor to perform certain items of work in the absence of total agreement with the contractor as to cost. If a change in contract sum is anticipated, a price adjustment will be made using one of the four methods:

- Mutual agreement on a lump-sum basis
- In accordance with units prices contained in the contract
- Costs determined in a manner agreed on by all parties
- Documented cost of the work, which will include
  a. Cost of labor to include all fringe benefits
  b. Cost of materials, supplies, and equipment
  c. Rental costs of machinery and equipment exclusive of hand tools
  d. Cost of bond premiums, insurance, permits, fees, and taxes
  e. Cost of supervision and field office personnel directly attributable to this change plus a reasonable fee for contractor overhead and profit

It appears that in most cases where a contractor's cost proposal is not accepted because of what is perceived to be too high a price, the construction change directive presents a fair alternative. This article dictates how a change order containing both adds and deducts is to be totaled. Any deducts are to be applied to adds before contractor overhead and profit are calculated.

Article 8 concerns time. Contract time is measured in calendar days, not working days, unless otherwise noted in the contract. If the contractor's progress is delayed by actions or inactions of the owner or architect, the provisions of this article are invoked in order to request an extension of contract time. Since the architect is the one to determine what will be a reasonable time extension, he may have difficulty remaining objective if he has been the cause of the delay.

Article 9 delineates payments and completion. Applications for payment are to be submitted by the contractor at least 10 days prior to the date established for progress payments. Payments for off-site storage of materials and equipment are to be approved in advance, and in writing. The latest edition of A201 contains a definition of substantial completion, a term that was not well defined previously. *Substantial completion* is determined to be reached when the work or a portion of the work is sufficiently complete so that the owner can occupy the building and use it as intended. This implies a contingent Certificate of Occupancy being issued by the building official when Notice of Completion is filed. When substantial completion has been attained, the architect is to prepare a certification to include the date of substantial completion and contain statements as to division of responsibility for shop drawings.

The contractor needs to be aware of this to best position oneself to collect damages because the contractor suffered payment losses due to an unjustified delay in the architect's processing of shop drawings.

Schedule Acceleration: How It Is Used

The legal term "acceleration" should be a part of every contractor's vocabulary. If an owner directs a general contractor to finish a project ahead of schedule or if the contractor's work has been delayed and the contractor is then ordered to complete the project in accordance with the original completion schedule, these instructions are known as a demand for acceleration.

As stated previously, there are two types of acceleration: actual and constructive. Actual acceleration occurs when the owner demands that the contractor complete the project ahead of the originally submitted completion schedule and constructive acceleration occurs when the contractor is delayed by some action, normally an involved change order, or by an owner or architect's delay in reaching a decision on a question posed by the contractor. If the contractor then requests a job extension and is not granted one and the owner subsequently demands that the contractor complete the project according to the original contract schedule, a condition of constructive acceleration has been created. By claiming the condition of constructive acceleration, a contractor can attempt to get monetary relief because of the owner's actions. The legal elements of acceleration are as follows:

1. There is an excusable delay which entitles the contractor to a time extension.
2. The contractor submits a written request to the owner for a time extension.
3. The request for the time extension is refused, and the contractor proceeds with the work.

4. The owner issues an order to accelerate performance to complete the project within the original time period.
5. The contractor proceeds with the work at an accelerated pace and keeps track of the costs involved in the speedup effort.
6. The contractor then notifies the owner of the intent to file a claim to recover the costs incurred in this accelerated effort.

These situations will arise, and the contractor should be aware of the steps that can be taken to obtain reimbursement of costs caused by undue demands of the owner.

Speeding Up the Process

The procedure described above is cumbersome and impractical from a time standpoint, and the architect can speed things up by issuing a construction change directive, AIA Form G7143, which authorizes the contractor to proceed with the work promptly with costs to be determined by one of several means so that a formal change order can be written at a later date. This form, however, still has to be signed by the architect and the owner before the work is officially authorized, so that even the architect's attempt to speed up the process will take some time.

The most common approach to authorizing change order work is the verbal commitment or directive. The architect sends a sketch to the contractor discussing a possible change in the work. The contractor in turn prepares a cost estimate for the work. This may involve the contractor's own cost estimates, including subcontractors' prices, or a combination of both. The contractor's allowable markup for overhead and profit will be added to the net cost, and the total price will be faxed to the architect or passed on verbally at a job meeting. The architect will consult with the owner and advise the general contractor whether the price is acceptable and the work should proceed or whether the price is not acceptable and arrive at a compromise scope and price so that the work can get started.

At this point, the general contractor should confirm the instructions he received, including the price, by sending a letter to the architect, unless the architect has already initiated a confirming letter. The contractor's confirming letter can be short and simple and should include the following:

"Pursuant to the issuance of a formal change order, as of this date, (name of contractor) is to proceed to (describe the nature of the extra work) based on your verbal authorization to do so. The cost of this work, as mutually agreed on, is $ and includes the general contractor's overhead and profit. It does not appear that this work will have any impact on the construction schedule (or if it will have an impact, state so here). It will be appreciated if a formal change order can be issued at your earliest convenience." The contractor should remember that although there has now been a verbal authorization to proceed followed by a written confirmation, it is still important to obtain a formal change order as rapidly as possible. The extra work may be completed promptly, but it cannot be included on the monthly requisition for payment until the formal change order has been duly executed by all parties. For that reason alone, it is wise to press for prompt processing of the formal change order.

There are certain facets of change order preparation that should be given more than just casual thought. The question of what constitutes allowable costs and overhead and profit percentages is not always clearly answered. Some contracts do not include specific lists of items constituting acceptable contractor "costs," and what may be applied against these costs for markup, and overhead and profit. Specific instructions and directions for change order procedures, if there are any, will usually be contained in the bid documents or the supplementary general conditions within the bid documents. By looking through Article 7 the contractor can see if there are any provisions for change order preparation that might have been overlooked. When preparing a cost estimate for change order work, the contractor should consider the following:

- Will the work contained in this change order affect the completion time?
- What is the net cost of the work? (The term net cost is explained below.)
- What other costs should be included in the estimate?
- Does the cost of this work have any effect on the payment and performance bond or on insurance requirements?
- What fees can be included in this change order work?

Change Order Completion Time

Here the questions that need to be asked concern activity duration. Are the nature and timing of the change order work apt to affect the completion date for this project? Will they extend the completion date, or will they actually decrease construction time? These questions should be asked even if a liquidated-damages clause is not part of the contract. There is a line item near the bottom of the change order form that pertains to contract time where the contractor is to indicate whether the contract time will be increased, be decreased, or remain unchanged by the issuance of the change order. A space is included to insert the number of days for the anticipated decrease or increase. Another line item directly below that one on the AIA change order form has a space in which to insert a revised date of substantial completion, if that will change.

Theoretically, an increase in contract completion time would mean that a request could be made for reimbursement of the portion of the general contractor's general requirements that is related to time. If the changes are going to extend the length of the job, the contractor could include the following as cost of the work: additional supervision, additional field office expenses (including office trailer rental), added temporary utility costs, added sanitary control costs, cleaning costs, and in some cases additional project management costs. If the nature of the change orders is such that it might be debatable whether the completion time will be extended, another approach might be tried. Get the architect to agree to the job completion extension time with the proviso that, if the project is complete within the original contract time frame, no claim will be made for additional general requirement cost reimbursement. If the completion extends beyond the date established in the contract or in the initial job progress schedule, there will be reimbursement of general requirements for the time extension agreed on.

One word of caution about requesting time extensions. Members of the project team can change during the life of a construction job, and although the contractor may have developed a fine working relationship with the original team, new faces can appear, and either the previously friendly relationships turn cool or go all the way to cold. If the architect or owner's representative had previously stated that formal time extensions were not necessary, because if the time came that they were needed, they would agree to them, the new team may completely disavow these verbal commitments. To avoid any such situation it is wise to prepare for the worst and request time extensions if they are justified. A contractor can always use the "hit-by-a-car" excuse not to imply that the owner/architect representative can't be trusted. Simply stated, if there is mutual agreement that a job extension is warranted, it should be so indicated in the written change order. Because in reality, suppose one of the team members *is* hit by a car tomorrow and there is no documentation of the verbal commitment?

Net Cost

Net cost is usually interpreted as the cost of materials; including sales tax and delivery charges; the cost of labor, including Social Security benefits, unemployment insurance taxes, fringe benefits established by collective bargaining agreements or company policies, Workers' Compensation insurance, bond premiums, and rental of tools and equipment; and fees levied by local, state, and federal agencies. Overhead costs are sometimes viewed differently by each company. Project management may be included in the estimate as a general condition expense, while other companies figure it as a corporate expense. The contractor's salary plus any other field-stationed time clerks are usually considered a general requirement cost; however, the cost of a secretary stationed in a field office may not be so easy to define. All expenses connected to the operation of the general contractor's main or branch office are historically defined as overhead costs, which also includes the top executives whose responsibility is overall management of the company.

Although some cost and overhead items are fairly clear-cut, the job superintendent's salary and related expenses might very well be a gray area. Depending on the nature of the anticipated change, the job superintendent may be required to spend more time supervising the change and may need an additional carpenter or labor supervisor to take over some other supervisory function temporarily. Obviously, a change order involving the installation of four more 120-V electrical outlets will not place an added burden on the superintendent. But what about the change in design of continuous footings and foundation walls to a system of spread footings, pilasters, and grade beams because of unsuitable subsurface soil conditions which have just been uncovered? And what about large quantities of interior partitions which are to be changed and will in turn require the relocating of a great deal of mechanical and electrical work already in place? Depending on the magnitude of the changes and the added demand placed on supervision, a strong case can be made for including additional supervisory costs in the cost of the work.

Don't overlook a cost sometimes referred to as "small tools." Will some hand tools need to be purchased to complete this change-order work? Generally hand tools are viewed as expendable; once used on the job, they have no further value. Will additional or different kinds of power tools be required to perform this work? Power tools are not considered expendable but are considered as having residual value after the job is complete, but a charge representing true depreciation of the tool or tools can and should be included.

Change Order Work Fees

The procedures and methods of costing, and the allowable overhead and profit percentage are normally established in the bid documents. Typical restrictions in these bid documents or supplementary general conditions may be termed as follows:

- The undersigned (general contractor) agrees that the total percentage for overhead and profit which can be added to the net cost of the work shall be___.
- This includes payment for work performed by his own forces or for subcontracted work, or for both.
- The undersigned (GC) also agrees that for subcontracted work the total overhead and profit percentages added by the subcontractor will not exceed the percentages indicated here ________.

In some cases, specific allowable overhead and profit percentages are stipulated for subcontracted work with these percentages decreasing as the cost of the work increases. When completing the bid document information, the general contractor is, in effect, establishing the overhead and profit percentages that a subcontractor will be allowed to charge for change order work. When subcontract agreements are being negotiated, after the construction contract has been signed, the subcontractor must be made aware of the restrictions that will be placed on the overhead and profit structure.

Along with the requirement for specific percentages to be applied for overhead and profit, the architect may also insert a clause that will affect the way overhead and profit are computed. A statement may be added that when changes in the work result in some cost increases and some cost decreases, the overhead and profit calculations will not be applied until the net change is established. The architect, by inserting the statement about overhead and profit being applied to the net change, would alter the request for payment indicated, and would dictate that the cost be tabulated differently:

The Construction Change Directive

Article 7 of the AIA General Conditions document contains a new term for change-order work: the construction change directive. This directive is an order signed by the architect and the owner directing the contractor to effect a change in the scope of work when there is an absence of agreement among all parties as to the terms of a change order, which usually means disagreement over the cost of the work. The cost of the work contained in the construction change directive is to be determined by a mutually accepted lump sum, by unit prices if contained in the contract, by "costs to be determined in a manner agreed upon by the parties" (whatever that means), or by a method prescribed in a subsequent paragraph. When contractor and architect-owner cannot agree on an acceptable cost for the work, the architect may request an itemized accounting of costs, which shall include the following:

- Cost of labor and related fringe benefits
- Cost of materials, supplies, and equipment, including cost to transport to the jobsite
- Rental costs of machinery and equipment, exclusive of hand tools whether these tools are rented from the contractor or other sources
- Costs of bond and insurance premiums, permits and fees, and taxes
- Additional costs of supervision and field office personnel directly attributable to the change order work

Approval Roadblocks

Invariably, the questions arise: Why is the change order process so fraught with problems and frustrations? After all, what could be simpler? The owner or architect would like to make a change in the work under contract, the nature of the work is discussed, a price is submitted and agreed on, and the work is completed and is paid for on the next requisition. Another change-order scenario has the general contractor uncovering a discrepancy, omission, or conflict in the contract documents and requesting that the architect review the problem and discuss the additional costs to correct the situation so that a change order can be issued. Seems simple and natural enough in a work process. But it couldn't be more different.

The greater portion of change order work is usually generated by the owner or architect, when a complete, well-coordinated set of plans and specifications has been the basis on which a construction contract was prepared. Most general contractors would prefer to go through a project without the disruptions and delays that can be caused by change order work. The exception to this rule is when a general contractor finds that the construction plans and specifications are deficient in that they do not fully cover the scope of the actual job requirements and such significant additional work is required that additional compensation is justified. This situation probably sets up the first roadblock in the relationship between contractor, architect, and owner.

The ability to view problems from the other person's perspective can be particularly important when it comes to change-order requests. Let's look at it from the general contractor's viewpoint first. A small or midsize general contractor may be estimating many jobs at the same time and only a certain amount of time is allotted to each estimating effort. The estimating department will be estimating exactly what is shown on the drawings, and there may not be time to analyze whether a certain detail will work. It is the general contractor's responsibility to price out what is on the drawings, and the responsibility for proper functioning is left up to the design professionals. If the general contractor is awarded the job and, on commencing construction, finds that the design documents are incomplete, why shouldn't the contractor be allowed to request the additional funds necessary to complete that portion of the work? The contractor must assume that the plans and specifications which were to be estimated represented a complete project that had been assembled and checked by design professionals for accuracy, completeness, and compliance with all applicable codes and regulations.

The Architect-Engineer Viewpoint

In the preparation of drawings and specifications for the construction of a sophisticated project, by constraints of time and money, it is almost impossible for the architect to include all the construction details. If the drawings are too exact, they might not give the contractor any latitude to adapt to field conditions. The architect recognizes these facts and establishes certain ground rules to cover some of the contingencies, but by their very nature they may also create some of the problems associated with change order work.

The first potential area of conflict between contractor and architect is addressed in the bidding documents. There are usually statements intended to discourage change orders. Instructions to bidder statements generally include a phrase that directs the contractor to notify the architect, in writing, of any ambiguities, inconsistencies, or errors in the bid documents, prior to submission of the bid.

The next warning is usually contained in the bid proposal, which states that the undersigned general contractor is to affirm by signature that all the contract documents have been examined, and all rights to plead any misunderstandings of these documents are hereby waived. From there we go to the "supplementary general conditions of the contract for construction," which will contain phrases to the effect that should the general contractor discover any contradictions, ambiguities, errors, or inconsistencies in or between any of the contract drawings, before proceeding with the work, the architect will be notified and requested to render an interpretation of these matters. Failing to do so, the contractor will have no excuse for not carrying out the work in a satisfactory manner as interpreted by the architect.

Still other supplementary conditions state that all drawings are a working part of the specifications and any questions or disagreements as to the true intent of the specifications or drawings, or the kind or quality of work required, shall be decided by the architect and/or local authority, whose interpretation shall be final, conclusive, and binding on all parties. Now this is very unilateral and nonnegotiable language the astute contractor will not accept. The architect, knowing that the drawings might lack some specific details, would like to be protected from that unscrupulous contractor who might be looking to take any and every opportunity to generate change orders. So the honest, hard-working contractor needs protection from that less-than-perfect design consultant. I don't mean that derogatorily—it's just the way it is.

So what is the solution? If it is impossible for the architect and engineer to issue a perfect set of plans and specifications, complete in every single aspect, it is also impossible for the general contractor, during the bidding process, to comply with all the investigations for completeness of design as required by the architect in the bid documents. It would appear that the best solution to the controversies that arise over change order work would be to require both contractor and architect to be reasonable in their approaches to these problems of extra cost items. Is it reasonable to assume that it is the contractor's responsibility to ensure that the mechanical and electrical drawings have been properly coordinated with each other and with the architectural drawings? If during construction it is determined that the ductwork, sprinklers, and light fixtures can't fit in the space above the proposed ceiling and additional costs are involved in rerouting some ductwork or piping, is this not the basis for an unarguable change order request? On the other hand, if additional window blocking is required for proper and secure window installation, and this blocking detail is not specifically shown on the drawings, is this a legitimate claim for an extra cost request?

The answer to the first question is yes. The contractor cannot and should not be held responsible for the coordination of multidisciplinary drawings that is necessary to make certain that everything will fit in an allotted space. The court has held that this does not seem a reasonable responsibility to be assumed by the general contractor. The answer to the second question is no. It does not seem reasonable for an architect to assume responsibility for the complete detailing of window blocking, since an experienced and competent general contractor ought to be able to develop the proper wood-blocking details as field conditions dictate. Both questions should be viewed with an eye to what is reasonable; and if that kind of approach can be developed, some of the difficulties in implementing and approving change-order work may be eliminated.

The contractor should also bear in mind that the architect does not relish having to go back to the owner and reveal that extra costs might be involved because of omissions on the drawings or details that lack clarity. There are usually compromise positions that all parties can take to resolve these problems. One approach that the contractor can take is that, if a detail or missing items had in fact been shown on the drawings at bid time, all the bids would probably have included additional costs for the items. Therefore, the initial project cost would have been higher. Since that was not the case, it is merely a question of including the costs now instead of having them in the initial bid.

Liquidated Damage Clauses

There are a few terms connected with change order claims process that may surface, particularly when there are problems getting change order work approved. One such term is *quantum merit*, and another is *unjust enrichment*, and they present the contractor with another approach to obtaining change order approval. The Latin term *quantum meruit* is also referred to as "quasicontract." Both terms refer to recovering costs for benefits received rather than from the contract between two parties. The theory of *quantum meruit*, in the legal community, is that damages or claims for money by a contractor in connection with such things as change order work will be based on the value of the work performed. If there is a dispute between owner and contractor over change order work and whether it was authorized, the contractor can base a request for payment on the quantity and quality of work performed or in question. The theory of unjust enrichment differs somewhat from the theory of *quantum meruit* in that unjust enrichment will be used to measure damages or contractor's claims by the amount by which the owner has been enriched. There may come a time when the project manager will proceed with change order work based on a good-faith commitment from either the architect or the owner. Let us assume that this oral agreement to do the extra work was never confirmed in a letter or memorandum. Now that the work is complete, the contractor is requesting that a change order be issued so that payment for the work can be requisitioned. Repeated requests go unanswered for some reason; and since there was no written authorization to do the work, the project manager is now concerned about being reimbursed for the cost of the work and the associated contractor's fees.

Let us assume that the change involved upgrading the birch hollow core doors, as specified by the contract documents, to oak veneer, solid core doors. There were 10 doors involved, and the additional cost per door amounted to \$150; therefore, the total cost of this extra work was \$1500 plus the contractor's fee. The owner, very clearly, has had the project's value enriched by this substitution, and if it looks as though the owner will not issue a change order for this work, the contractor should familiarize the owner and architect with the legal concept of unjust enrichment. The owner may argue about the costs associated with this change and request documentation to substantiate every aspect of the \$1500 cost, but the argument that value has not been added to the job by construction, however, they often require that various life-safety items be added or modified. If those items result in added costs, the owner must recognize that the costs are reimbursable to the contractor. The time to uncover and discuss the restrictive contract language referred to above is when the contract is being reviewed at the beginning of the job, not at the end. But contractors who are not willing to risk being argumentative at the beginning of a project often find themselves immersed in problems at the end. The liquidated-damages clause is a provision inserted into the construction contract stipulating that a certain monetary penalty will be leveled against the contractor for each day that the project completion is extended beyond the agreed date, taking into account mutually agreed-on time extensions. A typical liquidated-damages clause will usually be inserted into the paragraph referring to contract time in construction contracts. It will probably be similar to the following:

"If the work is not substantially completed in accordance with the drawings and specifications, including any authorized changes, by the date specified above, or by such date to which the contract time may be extended, the contract sum stated in Article (whichever article contains the contract sum) shall be reduced by $___ (the daily amount of damages) as liquidated damages for each day of delay until the date of substantial completion."

Although the intent is to have the contractor pay the owner for damages incurred due to late delivery of the building, the liquidated-damages clause states that the total amount of damages will, in actuality, be deducted from the final payment, since it has the effect of decreasing the contract sum. It was first thought that by having a liquidated-damages clause in a contract, the contractor would have more reason to complete a project on time. If the contractor didn't, the owner's damages represented by the daily dollar amount of liquidated damages would be compensation for late delivery without recourse to legal action. That has not always been the case, inasmuch as the courts have been used to interpret and determine what constitutes the proper application of the liquidated-damages clause. One of the key elements of the liquidated-damages claim is "complete, including any authorized changes." The question of contract time that can be extended by change orders has already been discussed, but what about the other justifiable reasons for extending contract completion time?

Job Delays

There are three basic categories of construction project delays in the eyes of the law:

- Excusable. The contractor is granted a time extension, but no monetary compensation.
- Concurrent. Delays occur, but neither contractor nor owner can collect monies for damages.
- Compensable. Delays for which either owner or contractor is entitled to additional monies.

Excusable Delays

Excusable delays will allow the contractor to extend the completion date of the contract but may not allow him to recover costs associated with the delays. The following constitute excusable delays:

1. Acts of God
2. Fires and other accidents

3. Illness or death of one or more of the contractors
4. Transportation delays over which the contractor has no control
5. Labor strikes or disputes
6. Unusually severe weather

Concurrent Delays

The contractor generally assumes some delays for weather conditions and for periods of labor nonavailability. When delays are concurrently caused by the failure of the owner to honor contractual obligations and the contractor's lack of performance, the courts have been prone to rule that each party must bear its own losses because of the delays incurred. In other words, if the owner is remiss in meeting the requisition payment schedule contained in the contract and the contractor arbitrarily, without any written notice, slows the job down because of the late payments, neither party will be able to collect on the damages attributable to their delays.

Compensable Delays

Compensable delays, according to the courts, are those delays for which damages can be claimed. Compensable delays include delays caused by elements beyond the control of the contractor, but within the control of the owner—such as changes in the work, access to the site, and site conditions differing materially from those specified in the contract. Specific court cases have established some delays as "compensable;" these are cited as follows.

- Delays in the approval of shop drawings submitted by the contractor. [Specialty Assembling and Packing Co. v. United States, 274, Ct. Cl. 153 (166).]
- Delays caused by owner's improper inspection procedures. [Gannon Co. v. United States, 189 Ct. Cl. 328 (1969).]
- Inadequate or defective drawings or specifications. [United States v. Spearin, 248 U.S. 132 (1918) and J. D. Hedin Construction Co. v. United States, 171 Ct.Cl.70 (1965).]
- Contract changes when the nature of the work changed affects the original or unchanged work and causes an extension of time. [Conduit and

Foundation Corp. v. State of New York, 425 N.Y.S. 2d 874 (App. Div. 1980).]

- Owner work-force interference. The owner, in accordance with the provisions of the contract, is allowed to award separate contracts for work on the project, and this work interferes with the general contractor's work. Such a situation could arise on a union construction job when an owner engages nonunion subcontractors and jurisdictional disputes lead to work stoppage. [Bateson Construction Co. v. United States, 319 F.2d 135 (Ct.Cl.1963).]

When do liquidated damages commence, and when do they cease? The date for commencement is stated in the contract, and this date can be extended only by agreed-on extensions of time and excusable delays. When do liquidated damages stop? They stop when the project reaches the stage of substantial completion as determined by the architect in accordance with the provisions of Article 9.8 of the AIA General Conditions document. If there is a disagreement between the contractor and the architect as to what constitutes substantial completion, the responsibility falls to the general contractor to establish that the project in contention was completed to the point at which it could be used for the purposes for which it was intended.

Some contracts may state that liquidated damages will remain in effect until the project is 100% complete and all current punch-list items are also complete. This provision may not be enforceable since the courts have generally gone along with the substantial completion concept.

Enforcement of the Clause

Liquidated-damages clauses will not be enforced by the courts if they are construed to be used as a penalty. Even though the language states that the liquidated-damages clause is not a penalty clause, that's immaterial. Certain criteria must be present for the clause to be acceptable:

1. The amount fixed as the daily dollar amount of damages must be a reasonable assessment of the costs the owner will incur if the project is not completed on time.
2. The effects of the breach of contract (inability to finish on time) must be very difficult to establish at the time of contract preparation.

The amount of the liquidated damages must bear a relation to the real, and not imagined, harm that may be caused to the owner if completion is not adhered to. In a case known as Harty v. Bye, 483 P.2d 458 (1971), the Supreme Court of Oregon ruled that, although a contract contained liquidated damages and the contractor did not complete the project on time, the owner was unable to substantiate any proof of actual losses due to the late completion. In another court case known as Nomellini Construction Company v. Department of Water Resources, 96 Cal. 682 (1971), neither party could establish the amount of damages which were incurred as a result of the other party's delays. The court said that if it is not able to discern a clear-cut distinction between the damages caused by the parties, then liquidated damages cannot be assessed.

A further illustration of unenforceability is the case of Mosler Safe Company v. Maiden Lane Safe Deposit Co., 199 N.Y. 479 (1910). There was a liquidated-damages clause in the construction contract, and when the contractor completed the work later than the date specified in the contract, payment was requested in full per the contract amount. The owner wanted to invoke liquidated damages. During the court action, the contractor stated that he had been delayed because the architect required the work to be installed in a manner different from that in the contract drawings. The contractor also claimed that the architect delayed the job by not acting promptly on shop drawings submitted for review. The court ruled that, when delays are caused by both parties, it will attempt to apportion the damages and will refuse to enforce the liquidated-damages provision of the contract.

Documentation Required for Liquidated Damages

With respect to excusable delays (delays which will extend the contract time but will not support a claim for additional monies) the contractor should submit a written request for job extensions whenever they occur, even though it appears at the time that the job is well ahead of schedule. When a request for job time extension is made, there may be no immediate response from the architect. If no response has been received after a week or so, it is a good idea to write another letter referring to the first request and stating that if there is no response to this second request within five working days after receipt of the request, it is assumed that the delays have been accepted and therefore the contract completion time will be extended by the requested number of days. This will trigger a fast response. When weather is the basis for requesting job delays, it is important that the job superintendent document the daily time sheet or daily diary properly. The report or diary should have an entry such as "No work day due to extremely heavy rains from 6:30 a.m. to 10:30 a.m." If any workers have reported for duty awaiting a decision on whether the day will be considered a no-work day, that also should be noted in the report. When a claim for weather delays is made, copies of the primary documents should be attached to the request.

Change Order Claims Histories

The problems involved with change order work has been discussed in other parts of the book, and the theories of *quantum meruit* and unjust enrichment have been just previously explained. A look at a few court decisions will shed additional light on the question of liability for payment for extra work not specifically authorized by written change orders. In a case known as Universal Builders Inc. v. Moon Motor Lodge, Inc., 430 Pa.550, 244, A.2d 10 (1968), the owner's agent made several requests to the contractor involving extra work. The contract contained language related to change orders and required that all change orders be in writing and be executed by the owner or the architect. The contractor did some extra work without written change orders but with the promise of the owner's agent to pay for the added work. After the work was done, the owner refused to pay, invoking the contract clause that all change order work must be in writing.

When the case was reviewed by the Pennsylvania Supreme Court, the ruling was that the owner's agent not only promised to pay for the extra work but also saw the contractor performing the extra work while visiting the site on several occasions. Rather than protest, the agent watched the extra work being incorporated into the job. The court interpreted this as an implied promise to pay for the extra work. The court also said that it would be unjust to let the owner accept the benefits of the extra work just because a written authorization did not exist. In the case of the GC and a subcontractor, the GC required the subcontractor to submit proposals for all extra work before proceeding with the work. The subcontractor proceeded to do some work without complying with the GC's requirement for written authorization and had to sue to recover their time and material costs. The court ruled that there was clear evidence that the GC was aware that the subcontractor was doing the work without written authorization. As a result, the subcontractor was able to recover their costs [Griffin u. Geneva Industries, 228 N.W., 2d 880 (1975)].

Computerized Evidence

The use of CPM as a method of pinpointing delays has been further enhanced by a change in the rules of evidence in the federal courts. In 1995, the federal courts adopted a new set of federal rules of evidence which filtered down to the state courts and has also been adapted to arbitration proceedings. Among the changes has been the recognition of the accuracy and reliability of computerized reports and summaries and the CPM schedule. Computer output is based on acceptable input quality. The AIA General Conditions article (Document A201) relating to contractor's computerized schedules has changed over the years. The 1970 edition required the contractor to submit its schedule for the architect's approval, while the 1976 version required submission for the architect's information. The 1987 version of A201 merely states that the contractor shall prepare and submit a schedule promptly. These distinctions, while subtle, are important. Schedules are dynamic and change quite frequently, and with the submission being for the architect's information, the GC has more latitude and flexibility in modifying the schedule and documenting the reasons to substantiate any changes that may affect job progress or project completion without being held accountable for the initial schedule submitted to the architect.

Claims Against Professionals

Architects and engineers, in recent years, have been subjected to an increase in lawsuits against them. The standard professional liability for malpractice insurance refers to "errors of commission" and "errors of omission," among other things. Now, developing a set of drawings and specifications for a major project is a very intricate and complex endeavor. Using the old phrase "Nobody is perfect," it is certainly understandable that the final construction documents might be lacking somewhat, and the GC routinely expects to have to add its expertise to the drawings to provide a completed project. With the drastically lowered cost of personal computers and computer-assisted-design (CAD) software, another element has been added to the equation. Construction details can be selected from an electronic brain and inserted into a building's design with the click of a mouse but without being guided by a hand experienced in how things fit together, the resultant drawings can have serious design deficiencies.

The contractor is expected to give its personal attention to the work, and either the contractor or its authorized representative must be on the construction site at all times that work is being performed there. Some contractors will leave the jobsite without adequate supervision in an attempt to entice the resident engineer or inspector into directing some of the construction work (actual direct supervision). Such action by the resident engineer or inspector would then shift a portion of the job responsibility to the owner or the architect/engineer. In cases of defects in construction in which the inspector has exercised such supervision, the contractor has grounds for being legally relieved of the financial responsibility of correcting such defects. In addition to the requirements for providing continuous, adequate supervision, the contractor is also required to conform to all laws and ordinances concerning job safety, licensing, employment of labor, sanitation, insurance, zoning, building codes, environmental conditions, and similar constraints.

The general contractor, as the only party with a direct contractual relationship with the owner for construction, is responsible for and must guarantee all materials and workmanship, whether constructed by its own forces or by its subcontractors. The owner should check the specifications thoroughly, as they may contain a provision that the contractor must arrange for an extension of the performance bond through the entire guarantee period, usually one year from the date of completion of the work and its acceptance by the owner. The requirement may call for the bond in the full amount of the original contract price or may allow a reduced amount such as five or ten % for the term of the guarantee period.

Even though the contractor normally has no direct responsibility for the plans and specifications, it can incur a contingent liability for proceeding with faulty work whose defects should be obvious to one in its business. Thus, if the contractor is asked to do something that it feels is improper or not in accordance with good construction practice, the contractor is entirely within its rights to submit a letter of protest to the architect/engineer or owner stating its position before proceeding with the matter in dispute. If after the contractor has submitted its written protest, the architect/engineer or owner prefers to require that the contractor proceed anyway, the contractor is protected in case of later flaws that may develop. It may even be possible that the insuring firm could get out of paying a claim if a flaw developed later as a direct result of not heeding the contractor's advice—a matter for the courts to decide. Insurance coverage is an important contractual responsibility of the contractor, both as to type of insurance and policy limits. The contractor is required to provide insurance not only for its own direct and contingent liability, but also frequently for the owner's protection. The contractor is expected to exercise every reasonable safeguard for the protection of persons and property in, on, and adjacent to the construction site.

Some of the most important of the contractor's rights include progress payments, recourse by the contractor if the owner fails to make payments, termination of the contract for cause, rights to extensions of time and extra payment as provided, and appeals from the decisions of the owner and the architect/engineer. Subject to contractual requirements and contractual limitations and, in the case of public works contracts, subject to the local laws concerning the use of subcontractors, the general contractor is normally free to purchase materials from any source it wishes, and to schedule work in any manner that it sees fit in the absence of any contractual provisions to the contrary.

Arbitration or Litigation?

Many standard forms of construction contracts include provisions for binding arbitration of any future disputes. It is also the declared public policy of the federal government and many states to require arbitration in public contracts. Yet we find that most insurance carriers try to avoid arbitration wherever possible. Parties to the construction process often gain a false sense of security from having an arbitration clause in their contracts. There is no perfect method for dispute resolution. Whether it is arbitration or litigation, the choice between the two is much like the choice of whether to be put to death in a gas chamber or in an electric chair. Under both arbitration and litigation the parties are asking an outside party to resolve their disputes. Common sense dictates that the quickest and most economical way of resolving a dispute is for the parties to resolve the dispute themselves, without the intervention of others.

Is arbitration the best method for resolving construction disputes? The answer is a definite "maybe." The proponents of arbitration hail it as being speedy and economical (see the section on Arbitration earlier in the book). Experience has shown, however, that this is not always true, especially in the larger cases. Arbitration can be extremely lengthy, expensive, and by its very nature can lead to inconsistent results. One arbitration case in which the author was involved lasted over five years mostly because of hearing delays and the difficulty of scheduling hearing dates that were mutually acceptable to the arbitrators, the parties to the arbitration, and their attorneys who also had to weave their arbitration schedule in and out of equally demanding court appearance dates, many of which occurred on short notice, thus forcing the arbitration to be rescheduled.

Chapter 14
Advanced Claims Strategy

Records and Defense

There is no claims protection or support without adequate documentation. The single most important thing that a contractor can do prior to the existence of a claim is to provide the attorney with the tools needed to defend or prosecute its case, whether it be in arbitration or litigation.

The construction environment of today is far too complex and vulnerable to claims to leave matters to be settled orally in the field. All matters of substance should always be in writing, and it is better to document too much than too little. If sufficient documentation is maintained, many potential claims problems can be resolved equitably without resorting to the courts. The same records that can assist in the resolution of difficulties in the field can be used to support your position if a claim later goes to litigation or arbitration.

Many of the records that are kept in the normal course of business are admissible as evidence in litigation or arbitration, so it is important that sufficient data be recorded and that the documents meet certain standards to be meaningful. Diaries can normally be used by a witness during testimony as an aid to his or her memory if such diaries meet the prescribed requirements for credibility. In some cases, such diaries are admissible as evidence in their own right as a record kept in the normal course of business.

Properly kept records are your first line of defense. Don't lose a case before it even starts by failure to develop and implement an effective record-keeping system. It is equally important that all parties working on a project be thoroughly indoctrinated in the proper use of the record-keeping system. The best system in the world is of no value if not faithfully used.

Contract Documents Precedence

To assist in the resolution of field disputes in cases where a discrepancy exists between the various parts of the contract documents, it is desirable to establish some system of order to indicate the relative order of importance of each. Although through legal precedent some degree of order has already been established, many General Conditions cite the specific order so that a dispute may be settled without the need for interpretation or arbitration.

Although a specific listing of order, if specified in the contract documents, will govern over any policy listed here, the following guide is based upon the NSPE General Conditions, with some modifications. The contract documents are complementary; what is called for in one is as binding as if called for in all. If the contractor finds a conflict, error, or discrepancy in the contract documents, it should be called to the attention of the engineer in writing before proceeding with the work affected thereby. In resolving such conflicts, errors, and discrepancies, the documents should be given preference in the following order:

1. Agreement
2. Specifications
3. Drawings

Within the preceding elements, further preference can be established as outlined in the following breakdown of a typical set of contract documents (on private contracts it is uncommon to have the Notice Inviting Bids, the Instructions to Bidders, and the Bid or Proposal as a part of the contract documents). Within the specifications, the order of precedence is as follows:

1. Change Orders or Work Change Directives
2. Agreement
3. Addenda
4. Contractor's Bid or Proposal
5. Supplementary General Conditions (Special Conditions)
6. Notice Inviting Bids
7. Instructions to Bidders
8. General Conditions of the Contract
9. Technical Specifications
10. Standard Specifications
11. Contract Drawings

With reference to the drawings, the order of precedence is as follows:
1. Figures govern over scaled dimensions
2. Detail drawings govern over general drawings
3. Change order drawings govern over contract drawings
4. Contract drawings govern over standard drawings
5. Contract drawings govern over shop drawings

As might be expected from a document prepared by the owner, the contractor's rights under the terms of the construction contract documents are often few and its obligations many. However, certain standard forms of General Conditions of the Construction Contract, such as those of the AIA, the EJCDC, and many public agencies, provide fairly equitable conditions for all parties to the contract, which is one reason that builders generally favor such standard forms over those typewritten, do-it-yourself varieties produced by the usually well-meaning but often poorly informed specifier or even attorney who is unfamiliar with construction contract requirements.

Invariably, every claim seems to include some form of estimate. An objective review of the source material and methods upon which the estimate was based will often show up weaknesses or deviations from accepted industry practice. In any case, it is extremely difficult to determine the amount of a claim or counterclaim without preparing careful estimates of the work involved. Another of the items of construction that seems to contribute to the most controversy is scheduling. Scheduling controversies arise not only from poor planning, on occasion, but also from disagreement over construction methods or sequences of operations. These disputes most often occur where a project is constructed under multiple prime contracts, such as those commonly experienced in a professional construction management (PCM) contract. The absence of any contractual relationship between the individual prime contractors or the architect/engineer has been the root cause of numerous disputes. Scheduling techniques such as CPM or other network scheduling systems and their associated computer-generated printouts have only served to compound the problem. Often the owner has stacks of computer printouts that it either does not want or does not know how to read. Yet somewhere in that stack of paper may well be the key to success in the defense of claims being made against it. The question of scheduling methods of operation and the impact of deviating from the anticipated schedule requires a careful analysis to determine the reasonableness of the originally anticipated schedule in conjunction with the planned methods of operation. The analysis of every schedule-related problem is unique, and every such analysis should include a review of the anticipated sequence and schedule together with a review of the actual progress of the work. The review must include an analysis of any delay and impact caused by all parties to the progress of the project.

Costs

It is in the presentation of cost data and their supporting documentation that most construction claims are the weakest. A carefully detailed reconstruction of all construction activities and their related costs must be made and presented so that they can be clearly identified for use in negotiating affirmatively or defensively. Construction-related organizations are reported to be poor in their cost record-keeping systems (or lack of systems) and a lack of ability to relate costs to possible claims areas. An accounting firm is not the answer either, as often those cost items relating to potential claims are not identified by an accounting firm because most accounting firms are not familiar with the cost control procedures for construction contracts.

Litigation Planning Strategy

Before a dispute even reaches the courtroom, the value of project records becomes quite evident. During the early stages of case preparation they can be used in helping define the issues of a complaint. They are essential to the establishment of the facts in a case and to the planning of litigation strategy. In addition, project records are valuable as an aid to weighing the credibility of an additional source. During the prelitigation phase of case preparation, the project records are invaluable in the preparation of interrogatories and depositions, for without competent and thorough documentation, key issues could easily be overlooked. They are also of considerable value during the preparation of the trial outline itself.

Initially, the project records may be used to define contractual parameters within which the parties should function. Then, during litigation, these records may be used to define relevance and establish probative value of facts under dispute. The value of the records does not end here. They are also used to establish credibility or, conversely, for impeachment. Project records are also of considerable value in ascertaining the true impact (damages) from the harm that was alleged or suffered by the general contractor, subcontractors, and/or suppliers. A contractor does not waive its rights to claim extra costs over and above the amount allowed by force account, if it can be justified and verified; however, it is contractually bound to proceed with the extra work as long as the owner continues payments to the contractor under the terms of the change order. It should be kept in mind that a force account settlement is open-ended as opposed to a lump-sum settlement which may cover all costs of an item of extra work.

Another important source of evidence is the live witness, including the expert witness. Contractors should provide their attorneys with a list of all potential witnesses, including those who may testify against the claim, since it is equally important that the opposition's strengths are known early in the case. A live witness is especially important where the dispute concerns an oral agreement, such as an oral change order, for which there is no subsequent written memorandum. The exact spoken words can also provide valuable evidence where the dispute involves some subject matter that was left out of the written memorandum, either by error or because it was not directly relevant to the agreement. Given a solid reputation, the most important qualification of a person who will serve as an expert witness is integrity. The soundness of the integrity of the expert witness will be tested by the opposition in many ways, but most importantly by questioning the witness as to whether or not he or she is too closely connected with the case to give an impartial opinion.

In addition to proving the defendant's liability, the plaintiff is faced with the task of establishing the validity of the damages claimed. The general rule is that the prevailing party may recover whatever damages that may reasonably be supposed within the contemplation of the parties at the time that the contract was made. This is the most complex and time-consuming phase of construction litigation.

Often, the case is split into two separate trials, one for determining liability and the other for determining the damages. The advantages of separating the trial into two separate actions are that the determination of liability issues may narrow and simplify preparation for the damages phase; that it may enable the parties to get an earlier determination of the entire dispute by settlement after the liability issue has been decided; and that it may reduce the number of parties involved in the damages phase. The greatest disadvantage is that separation into two separate actions may delay final resolution.

Burden of Proof

Generally speaking, it is the plaintiff who has the burden to prove its claim. Where the contractor succeeds in proving substantial compliance against the owner's claim of nonperformance, the burden of proof reverts to the owner to prove its claim for residual defects. The ordinary fact finder, whether it be a jury panel or a judge, is not likely to be familiar with the technical and administrative details of construction. Therefore, the factual issues must be presented in a simple, concise manner. Photographic evidence and charts are very effective in conveying the facts. Actual site visits may save time and energy consumed in otherwise describing the general physical conditions. An expert witness who can translate technical concepts into simple language is an invaluable evidentiary aid.

Evidence

In all disputes, regardless of the method of resolution, the prevailing party will be the one that can better support its burden of proof of its allegation. In preparing to bring a dispute to some kind of resolution, the contractor must assure that it is equipped with all of the necessary evidence in support of its claim or defense, and anticipate what the opponent will produce as evidence of their position on the claim. The following is a checklist of the possible sources of evidence that should be used in the course of dispute settlement:

Documentation

1. Bid documents
2. Boring logs and soils reports
3. Drawings: as-planned
4. Drawings: as-built
5. Specifications
6. General and supplemental general conditions
7. Schedule: as-planned (CPM, etc.)
8. Schedule: as-built (progress charts)
9. Addenda
10. Change orders
11. Instructions and directives
12. Inspection records

13. Contractor's logs
14. Architect/engineer's or owner's diaries (through deposition)
15. Correspondence files
16. Check registers
17. Purchase orders
18. Shipping and delivery tickets
19. Time cards
20. Person-loading charts
21. Memoranda (including memos to file)
22. Site photographs
23. Testing results
24. Public records

The case will be constructed from public records, contract documents, and witnesses. The data required to support construction claims cannot generally be reconstructed after the fact without careful documentation. One of the most important functions of the claims-conscious contractor is the establishment and implementation of procedures for documenting what may eventually become a claim. Checklists should be made that can be used to check present documentation policies for their potential usefulness in the defense of claims. An owner would be wise to have its legal counsel investigate any engineering problem that might be likely to lead to a claim. Although there are instances where the architect/engineer can function effectively without the aid of legal counsel by making engineering determinations and documenting them for future use by legal counsel, there are very few instances where the attorneys should be involved in a potential construction claim without the advantage of the objective input of the architect/engineer. If negotiation fails, the documented data may be of considerable later value in litigation or in arbitration. Failure of a contractor to keep a diary or other daily record places that contractor in an indefensible position, and the contractor will be persuaded by its legal counsel to accept whatever out-of-court settlement could be reached, as the client has no case at all if they went to court.

Organized documentation of the facts surrounding the alleged wrongdoing is the single most important weapon in a case. There is nothing that can replace regular, detailed record keeping during construction. The tremendous volume of documentation is common to all projects, good and bad; therefore, the key is in the organization of the various documents. The contractor will have to provide or assist the attorney in producing a chronologically organized summary of facts of the dispute. At an early stage, the attorney should obtain copies of documents and other information held by the contractor through a pretrial process of discovery before they can be disturbed.

Claims Administration

There are several common forms of settlement of contractor claims costs. Principally, these include lump-sum settlements with or without qualification, and time and materials settlements. The type of settlement generally favored under larger projects is the lump-sum approach. With that type of settlement the owner and the contractor reach an agreement on the total amount of money and time required to complete the work. This agreement may be reached either prior to or after the work has been completed, with the advantage that it places a price ceiling on the entire dispute, and facilitates the grantee's application for funding from the construction funding lender. The most desirable form of a lump-sum settlement is the so-called "no fault" settlement, which allows resolution of the dispute without determination of fault. With this type of settlement, in exchange for the issuance of a change order, the contractor gives a full accord and satisfaction and a waiver of all claims arising out of the incident. A variation of the no fault settlement is one wherein the owner may obtain a unilateral reservation of legal rights that will enable the owner to sue for a portion of the claim settlement after full payment has been made to the contractor upon completion of the project. This type of solution is of special value by helping to solve the contractor's cash flow problems related to the claim and gives the owner a chance to recover funds that may later be determined to be ineligible for payment by the administrators of the loan funds.

With the time and material settlements, the contractor performs the work and submits documented charges for all materials, labor, and equipment used on the affected work. This type of settlement has one advantage, namely, the opportunity for the contractor to begin work without prior negotiation of costs involved. It has a disadvantage as well. By this method a contractor can artificially increase the cost and time required to complete the work and inflate the costs submitted to the owner for payment. A similar approach is sometimes referred to as *force account*. This is based upon a contractual agreement whose terms are generally spelled out in the General Conditions of the Contract for payment of extra work. It is generally applied in conditions involving an unexpected condition that leads to extra work and involves a prearranged formula for reimbursing the contractor for the costs of labor, materials, and equipment plus a fixed percentage for overhead expenses and profit, and the allowable surcharge that will be permitted to be applied to the billings of its prime and subcontractors.

Unforeseen Conditions

There is no uniform method of handling such unforeseen conditions. Numerous books and technical articles have been written about it, many cases are brought to suit over such differences, and the matter is no closer to the development of a fair and equitable contract provision that will apply to all such cases than it was at the beginning. Although their provisions may vary, construction contracts typically require the contractor to notify the architect/engineer immediately upon recognizing a situation that can lead to a claim or dispute. In cases of conflicts between the drawings and the specifications (a frequent cause of difficulty), it is the obligation of the contractor to notify the architect/engineer immediately without going further with what he or she knows to be improper work. There is always the possibility that a contractor who is alert for extras may recognize a defect in the plans or specifications, yet hastily construct the work exactly as shown on the defective documents, knowing all the time that it will be directed by change order to remove such work and reconstruct it later as soon as the architect/engineer or owner discovers the same defect. A situation such as this could be a race between the contractor and the inspector to see if the contractor can get the incorrectly specified or detailed work constructed before the inspector notices it and reports it to the architect/engineer or owner.

The logic of the plan is simple. If a contractor can feign ignorance of the problem and simply claim that it was just following the plans and specifications (which, undoubtedly, it would be) then, when the architect/engineer or owner discovers the defect in the plans and specifications and orders the newly constructed work to be removed and reconstructed, the contractor will willingly rebuild the work in any manner subsequently ordered by the architect/engineer or owner—followed later by claims for extra work, which, of course, the owner then becomes obligated to pay. If a dispute cannot be settled with the architect/engineer or owner, and they order the contractor to continue with the work as directed by the architect/engineer or owner, the contractor cannot usually refuse without becoming potentially liable for breach of contract. Although the contractor may perform such work under written protest, it must continue to do the work and keep the operation on schedule, relying on the contractual remedies in the contract documents to settle compensation time.

Resolution by Negotiation

Although the individual provisions may vary, most construction contracts contain language regarding the method of resolution of claims, either by litigation or arbitration. In the construction industry, there appears to be far more emphasis on the resolution of disputes by arbitration. Unfortunately, in a contract of adhesion the contractor is not offered a choice, except in those states where arbitration statutes prohibit contract provisions that agree at the time of signing the contract to submit to binding arbitration on future disputes.

Yet, in many cases, neither arbitration nor litigation is the sensible answer. Keep in mind that if the owner or architect/engineer goes to court or to arbitration, the costs on smaller claims may exceed the settlement, even if they win: whenever the value of a claim is under $30,000 or even $50,000, it is doubtful whether either party can ever win financially. The only true winners will be the lawyers and claims consultants. Fighting for principles may be fine for some, but to most people and businesses it is financially unsound. The only economically sound solution is to negotiate.

Negotiation involves compromise, so it should be entered into with that in mind. Remember that claims in the dollar range mentioned previously cannot ever be won by anyone except the lawyers, so even if a party collected only ten cents on the dollar, they would actually be ahead financially. Put up a good fight; do some hard bargaining, but be prepared to compromise. It should also not be forgotten that unless a little is left on the table for the other party to save face, they may be forced into an arbitration or litigation position that neither party can afford. While it might be good contract administration on the part of the architect/owner to attempt to resolve all matters relating to a contract modification [change order] during the negotiation of the modification, use of a clause which imposes an obligation on the contractor to submit a price breakdown required to "cover all work involved in modification" cannot be used to deprive the contractor from its right to file claims. This is an often misunderstood area in contractor claims. However, the probability of recovery by the contractor as the direct result of such conflicts is good insofar as the settlement is limited to the cost difference between the project cost as the plans and specifications are interpreted by the owner or architect/engineer and the contractor. A contractor should never attempt to construct any questionable area without first submitting to the owner or architect/engineer for clarification or notifying them of an error. In many contract forms, failure to do this may serve as a bar to full recovery of contractor costs.

Miscellaneous Problems

Problems may arise that are directly related to the conduct of the owner's representative on the job, such as issuing changes in the work that are of such magnitude as to constitute a cardinal change (creating a breach of contract) or indirectly related, where one subcontractor may negligently delay another, which may result in the owner seeking recovery from the negligent subcontractor in order to pay the subcontractor that was harmed. Some of the types of problems that may fall into this category include the following:

- Damage to work by other prime contractors
- Breach of contract
- Cardinal changes
- Work beyond contract scope
- Beneficial use of the entire project before completion
- Partial utilization of the project before completion
- Owner's nondisclosure of site-related information
- Owner's failure to make payment when due

Even if a perfect set of contract documents were to be devised, there would still be disagreements between parties. Disagreements may arise between the contractor and the architect/engineer or owner concerning the interpretation of the contract; what constitutes extra work on the contract; payment for changes; extensions of time; damages for owner-directed acceleration or slowdown; costs occasioned by owner-caused delay; defective drawings or specifications; and changed conditions (sometimes referred to as unforeseen conditions, based upon the logic that the conditions have not actually "changed" but were that way all the time; discovery of the difference between the actual conditions and the way they were represented in the specifications or drawings is referred to under this concept as "unforeseen" differences). Similar matters may affect contract cost or time required to complete the work. Contract documents routinely include procedures to be followed in the settlement of such claims and disputes.

Productivity Losses

In preparing a claim for acceleration or deceleration, it should be borne in mind that the costs to the contractor for going into premium time, such as working an extended workweek, cannot be computed simply as including the added hourly costs multiplied by the additional hours. Studies have shown that as the workweek is extended, there is an accompanying loss in worker productivity. Furthermore, as the extended overtime is continued, the productivity rate continues to drop.

Scheduled overtime is not often seen on competitively bid lump-sum contracts, as most contractors are well aware of the negative effects of overtime on cost and productivity. Simple arithmetic shows that premium pay for double time or time and one-half makes overtime work much more expensive. However, people who insist on overtime seldom realize that other costs associated with overtime may be even more significant than premium pay. Premium pay affects only overtime hours, but continuing of scheduled overtime drastically affects costs of all hours. All available research findings indicate a serious inverse ratio between the amount and duration of scheduled construction overtime and the labor productivity achieved during both regular and overtime hours.

In the first few weeks of scheduled overtime, total productivity per person is normally greater than in a standard 40-hour week, but not as much more than the number of additional work hours. After seven to nine consecutive 50- or 60-hour weeks, the total weekly productivity is likely to be no more than that attainable by the same work force in a 40-hour week. Productivity will continue to diminish as the overtime schedule continues. After another 8 weeks or so of scheduled overtime, the substandard productivity of later weeks can be expected to cancel out the costly gains in early weeks of the overtime schedule, so that the total work accomplished during the entire period over which weekly overtime was worked will be no greater, or possibly even less, than if no overtime had been worked at all.

When the loss of productivity is added to the higher wage cost (including premium pay), productive value per wage dollar paid after several weeks of scheduled overtime drops to less than 75% for five 10-hour days, less than 62% for six 10-hour days, and less than 40% for seven 12-hour days. When an overtime schedule is discontinued, it has been found that there was a dramatic jump in productivity per hour after return to a 40-hour week. Construction delay claims involving acceleration of the work usually include claims for loss of productivity, which often exceed all other claimed amounts. There tends to be a relative magnitude of the claim for loss of productivity.

Failure to Agree on Change Order Pricing

One of the most common causes of contractor claims occurs during attempts to price change orders. All too often, owner change orders contain a waiver clause that requires the contractor to guarantee that the price and time named in each individual change order represents the total cost to the owner for that change, and the contractor waives any rights to impact costs.

This, unfortunately, leaves the contractor with only one recourse: the claims process. The contractor should take note of court's settlement ordered in at least one contract dispute. In the Appeal of Centex Construction Co., Inc. [ASBCA 26830 (April 29, 1983)] the appeals court ruled that the owner may not force a contractor to "forward price" its impact costs when pricing an owner-directed change. This also comes into effect with acceleration. Acceleration of the work is usually the result of an attempt by the contractor to take whatever means and measures are necessary to complete the work sooner than would normally be expected for a given project under stated conditions, or an attempt by the contractor to take extra measures to make up for delays, whatever the cause, by utilizing whatever means are at its disposal to accomplish the objective. As stated previously, there are two types of acceleration: directed acceleration and constructive acceleration. They are largely self-defining, but briefly in review, a "directed acceleration" occurs when the owner or architect/engineer specifically orders a contractor to speed up the work. The U.S. District Court outlined the necessary elements of a claim for constructive acceleration. "Constructive acceleration is present when (1) the contractor encountered an excusable delay entitling it to a time extension; (2) the contractor requested an extension; (3) the request was refused; (4) the contractor was ordered to accelerate its work, that, is to finish the project as scheduled despite the excused delays; and (5) the contractor actually accelerated the work."

In considering acceleration, the first consideration must be whether the so-called "acceleration" is the result of an order from the owner to a contractor who is behind schedule to get back onto schedule, or an order from the owner or architect/ engineer to the contractor requiring a contractor either directly or constructively to complete the work prior to the scheduled completion date. If a contractor is behind schedule and it is the desire of the owner or architect/ engineer to require the contractor to get back on schedule, it is sometimes necessary to direct the contractor to take whatever means are necessary to assure completion by the originally scheduled completion date.

Herein lies an administrative risk for the owner or architect. If an order is issued to the contractor directing it to accelerate the work to catch up or make up for delays or lost time, the risk is that the contractor will perform as directed, then submit a claim for directed acceleration, arguing that although it appeared from the original schedule that it would not complete on time, actually the contractor may claim that there was no risk of failing to complete on time at all. The contractor might state that in the absence of the acceleration order it would have finished on time anyway.

The proper handling of a situation such as this is not to issue an acceleration order (even though acceleration is justified or needed), but to send a letter to the contractor calling attention to the fact that completion by the scheduled date will be required, and that "according to the schedule it appears that the contractor will be unable to complete the project by the completion date indicated in the contract. Please resubmit a revised schedule showing how the contractor plans to complete the project by the scheduled date." This leaves the means and methods to the contractor, as it should be by contract.

Temporary Facilities and Controls

These are the utility services that are essential to the construction process but do not form a part of the finished project. Under the Construction Specifications Institute (CSI) format, these requirements are spelled out in Division 1, General Requirements. Under other specifications formats this may not be as well defined. This is often a gray area, requiring clear definitions of responsibility at the time of the preconstruction conference. Some of the questions that must be answered on this subject are:

1. Who provides the services?
2. Who maintains them?

3. Who pays for these services?
4. If shared cost, in what proportion?
5. What are the contractual responsibilities, if any, under OSHA?

These issues cover:

- Storage facilities and staging areas
- Jobsite security during nonworking hours. Losses from vandalism and theft at unguarded construction sites are rapidly mounting sources of expense to the entire construction industry. An agreement sharing the costs of better security measures might be worthy of discussion.
- Cleanup and trash removal. Consider dumpster containers, scraps, sanitary wastes, and so on.
- Available hoisting facilities. Who supplies hoisting facilities? If the general contractor supplies them, what will be the arrangement to make the hoist available to individual subcontractors?
- Percentages for overhead and profit to be applied to change orders. What costs will or will not be included in the change order price?
- Length of time that a change order proposal price is to be considered firm.
- Identify individuals who are authorized to approve change orders.
- Procedures to be followed when submitting initiator change orders or change order proposals.
- Change order forms that must be used.
- Time extension requests made by subcontractors due to changes in drawings or specifications.
- Amount of detail required of subcontractors when submitting change order proposals or initiator change orders. Will a complete breakdown of all costs be required? Brief description? Descriptive drawings?
- Overtime resulting from change orders; consideration of decreased productivity.
- When materials or equipment are to be removed because of a change, which party owns the removed material or equipment, and who removes it from the site of the job?
- Responsibility for preparation of Record Drawings brought about by change orders.
- Warranty requirements.
- Employment practices.
- Listing and identification of all tiers of subcontractors.

- Punch lists. Establish timely punch list items and avoid an excessive number of punch lists.
- Record drawings and final document submittals.
- Final payment and retainage.

The establishment of meaningful communications between the parties involved on a construction project is essential for its successful development. Only with such a basis of understanding can the necessary planning and work proceed without conflict or costly disputes. The preconstruction conference is the logical means toward this end. (See "Precon" earlier in the book.) Obviously, no preconstruction agenda will contain all of the listed items above, but with the list as a reference while compiling the actual agenda, there is less chance that an item will be omitted.

Warranties

The term *implied warranties* means that the goods must be capable of passing in trade under the contract description and are fit for the purpose intended. *Express warranties* are those that are specifically set forth in the contract itself; they are in common use for many construction contracts. Warranties are generally for packaged items such as water heaters and compressors. Where the industry practice is to furnish a warranty for an item, the requirement for such a warranty may be included in the specifications. An express warranty is a means of achieving good procurement results by making the contractor responsible for its work and for failures of its work during some part of its useful life. The primary disadvantage of warranties is that they are often unenforceable. Moreover, a warranty clause costs money in the form of higher bid prices, and it cannot be demonstrated that the owner recovers the cost of warranty.

There are several obstacles to the strict enforcement of warranties. Even in what appears to be a clear-cut contractor responsibility under a warranty, certain action on the part of the owner or design firm may cloud the issue and result in litigation. After acceptance of an item by the owner, the operation and maintenance of the item is performed by other than the contractor's or manufacturer's personnel. Thus, many defects that occur during the warranty period can be argued to result from faulty operation and maintenance by others rather than from a defect in the item itself.

Additionally, the industry is becoming reluctant to accept several of the warranty clauses now in use. In particular, it will not accept the provision relating to third-party damages and the responsibility for an entire building and its contents, including damages to personnel, resulting from the failure of a single item. Warranties of sole-source items are also generally unenforceable.

In practical application, warranties are generally enforceable only as to defects existing at the time of delivery and acceptance. Although the warranty clauses usually require the contractor to obtain and enforce warranties normally furnished by manufacturers and suppliers, the exact nature of the warranty is usually not stipulated in the specifications. Therefore, their effectiveness is subject to the wording used by the guarantor and can be expected to be something less than specific. Such a warranty depends almost entirely upon the integrity of the guarantor (the manufacturer or supplier).

There are substantial differences in the warranty requirements of public and private agencies in their construction contracts. They range from requiring the contractor to warrant all work as to materials, workmanship, and contractor's design, to requiring the contractor to warrant only mechanical and electrical work. One agency may require the contractor to remedy all damage to equipment, site, buildings, and their contents resulting from a defect, whereas another agency may require the contractor to act as the owner's agent and obtain and enforce the subcontractors' and suppliers' warranties.

Generally, it may be said that the ownership of all materials used in the work or stored at the site is vested in the contractor until final acceptance of the work by the owner at the end of the job. Thus, any risk associated with the protection of the work and the repair of damaged work, or the delivery of damaged materials, is usually the responsibility of the contractor, not the owner. Similarly, the contractor's insurance carrier will normally be called upon to pay the costs of any such claims.

Delivery and Storage of Materials

There is no firm formula for the determination of which facilities will be accorded the contractor as a working or staging area or for its storage of construction materials. Generally, it is the contractor's responsibility to make its own arrangements for such facilities if provisions have not been made by the owner. Although it is true that most owners do provide space, in some areas no such space is under the owner's direct control; in such cases the contractor must make arrangements for space. Occasionally, this is done by the contractor actually entering into a rental agreement for space, which must accordingly be taken into consideration at the time of figuring the bid.

In no case should any contractor assume that it has the right to block public thoroughfares or to use public property of any kind, including parking lots, for its construction purposes, even when the work is being performed for the owner of the property under consideration, unless specific written authority has been granted. As a means of protecting from such claims, the contractor should be assured that the space being used by the contractor for its work area has been properly granted by the owner of the affected property.

Handling of Materials

The contractor must be concerned not only with the quality of the materials as delivered and their methods of installation, but should require that all such materials be properly handled during delivery, unloading, transporting, storage, and installation so that undue stresses will not result in latent defects that will not be detected until after the project has been signed-off.

When in doubt, the contractor should have the design firm or the owner contact the manufacturer of the affected material, who is generally just as interested in its proper handling in the field as is the owner. This is because the manufacturer is often the victim of unjust claims for defective materials when, in fact, the problem may have been due to improper handling during construction.

Contract Modifications

A change order is a written agreement to modify, add to, or otherwise alter the work from that set forth in the contract documents at the time of opening bids, provided that such alteration can be considered to be within the scope of the original project; otherwise, a contract modification may be required. It is the only legal means available to change the contract provisions after the award of the contract. Functionally, a change order accomplishes after execution of the agreement what the specifications addenda does prior to bid opening, except that an accompanying price change may be involved in a change order.

A price change is not necessarily always in the contractor's favor, however, as it could also be in the form of a cash credit to the owner, or it may involve no price change at all. It is standard practice in construction contracts to allow the owner the right to make changes in the work after the contract has been signed and during the construction period. Depending upon the contract and its specific terms, such changes might involve additions to or deletions from the work, changes in the methods of construction or manner of work performance, changes in owner-furnished materials or facilities, or even changes in the contract time or order of the work. Changes may also be executed to correct errors in the plans or specifications, or they may be the direct result of contractor suggestions that are approved by the owner and the architect/engineer.

Impact Costs

Often, an owner or architect/engineer fails to realize that the cost of changes in the work may well exceed the cost of the immediate change itself. Many change order forms contain an exculpatory clause that precludes a contractor from recovery of impact costs. In some cases, the two parties already agree on the price of a change in both time and dollars, but the contractor wants to reserve the right to file for impact costs. If this is all that the contractor wants, and all of the other terms of the document are acceptable, it may be advisable to accept the change order with that reservation. If the owner does not, the contractor can file a claim anyway, but the amount of the new claim will not only be the cost of any potential impact costs, but will include the now-inflated value of the time and dollars that had originally been agreed to.

A contractor would be ill-advised to sign-off on a change order if it expects to claim any future impact costs. A California court in Vanlar Construction, Inc. v. County of Los Angeles, 217 Cal. Rptr.53 (Cal. App. 1985) ruled that a contractor who signed off 81 change orders and seven supplemental agreements had waived its right to any impact costs arising out of the cumulative effect of the numerous changes. During the course of construction, the project issued 81 change orders and seven supplemental agreements. Each of these documents stated that it covered "all charges direct or indirect arising out of this additional work." Vanlar officials signed-off on each document. After completion of the work, Vanlar brought a claim for the impact costs that were not included in the individual change orders. Vanlar argued that it was impossible to measure the cumulative impact of the changes as each change order was issued, making it necessary to wait until completion of the construction. Vanlar also stated that it was a well-recognized custom of the industry to wait until project completion to submit a claim for impact costs.

The California court ruled that Vanlar's execution of the change orders and supplemental agreements precluded subsequent claims for increased costs arising out of those changes. The documents clearly stated that they covered indirect costs arising out of the extra work. The court also said that because the change orders were clear and unambiguous, Vanlar could not resort to industry custom to explain the intended meaning of those documents. In the court's words: "If Vanlar contemplated any future claim for impact costs, whether at the time of the first, fiftieth, or the eighty-first change order, it was legally obligated to request that a suitable reservation clause be inserted in each change order and supplemental agreement prior to accepting and signing the change order or supplemental agreement." If a project owner refuses to agree to appropriate reservation language, the contractor may perform the changed work under protest without signing the change order or agreeing to a price.

Bilateral Change Orders

The term change order, as normally referred to in all except federal contracts, refers to a bilateral agreement between the owner and the contractor to effect a change in the terms of the contract. In federal contracts, this document is called a contract amendment, as the term change order in federal contracts refers to a unilateral order to effect a change.

Unilateral Change Orders

A unilateral change in nonfederal contracts is referred to in the EJCDC contract documents as a Work Change Directive and in AIA contract documents as a Construction Change Directive. In federal contracts, a unilateral change is referred to as a change order. A unilateral contract modification is intended to expedite issuance of a change order to perform emergency work or protested work and must be replaced by a regular bilateral change order that addresses the effect of the change on contract cost and time before payment can be made to the contractor. When approved by the owner, the provisions of a change order become a part of the contract. If an approved change order is executed by the contractor, all of its provisions and terms are as binding upon the parties as are those of the original contract. Each change order must be evaluated individually; however, there are certain basic principles that apply to the handling of all types of change orders.

1. No work should be included beyond the scope of the base contract.
2. The identity of the individuals authorized to request and approve change orders should be established early. Such information should be provided to the contractor.
3. During the preconstruction conference, a meeting should be held to establish the change-order-handling procedures.
4. All changes in the work must be authorized in writing prior to the execution of any change.
5. The scope of a change order must be clear, and a request for a change order proposal should contain enough information to enable the contractor to make a realistic estimate.
6. The contractor should submit its proposal to execute a change order as soon as possible after receiving the request, and the owner's approval or rejection should follow as soon as possible.
7. The proposal should be fair. It should recognize the contractor's right to include overhead and profit percentages in change order estimates and in time and material change order billings. It should also be recognized that the contractor is entitled to compensation for legitimate time-delay claims, processing of deduct change orders, costs of disposing of removed material, and all other legitimate costs incurred in the execution of the change.

These factors are merely the first stage in the implementation of a contract change. This does not constitute a change order. During final stages, the change order form may be issued by the contractor in the form of a letter proposal and must address effect on time and cost. A proposed contract modification is announced until the matter is finally rejected or approved as a change order. If the procedures discussed in this chapter are followed, it is possible to minimize the inequities otherwise possible in the administration of a contract change order.

In some cases a situation may arise at the construction site where immediate instruction must be given, and revised drawings and specifications may be unnecessary. In such instances, the contractor, upon receiving proper authorization, should prepare an Initiator Change Order Request with instructions as to whether the work should proceed immediately or wait until the cost proposals have been approved and a formal contract change order issued. Whereas it is possible, under emergency conditions, to start such extra work under a Work Change Directive, it must be emphasized that any work performed beyond the scope of the original contract provisions must be followed up with a confirming change order clearly citing the effect of the change upon both project cost and time of completion.

In no case should any work beyond the original contract scope or requirements be executed solely upon the authority of a Field Order or Work Change Directive. All such work must be supported by a change order. This is for the protection of both the owner and the contractor. The owner has the obligation to review promptly any change order proposals and recommendations and direct the architect/engineer to reject or authorize such change. Change orders frequently have an impact upon the performance of other work that is not in itself changed. The term impact refers to the indirect delay or interference that a change on one phase of the work may create on another phase, and the costs of such delay or interference should be recognized as a consequential cost to be considered as a part of direct cost expenses of the contractor, and consideration must be considered for allowing for payment of these costs to the contractor.

Directed versus Constructive Change

Before addressing the sophisticated elements of change order prospecting in the contract, it will be useful to remember the two basic types of change: directed and constructive. A directed change is easy to identify. The owner directs the contractor to perform work that differs from that specified in the contract or is an addition to the work specified. A directed change may also be deductive in nature; that is, it may reduce the scope of work called for in the contract. In the case of a directed change, there is no question that a change occurred. Disagreements tend to center on questions of financial compensation and the effect of the change on the construction schedule.

A constructive change is an informal act authorizing or directing a modification to the contract caused by an act or failure to act. In contrast to the mutually recognized need for a change, certain acts or failure to act by the owner that increase the contractor's cost and/or time of performance may also be considered grounds for a change order. This is termed a constructive change and must be claimed in writing by the contractor within the time specified in the contract documents in order to be considered. The owner should evaluate a change order proposal based on such a claim and can use the same reasoning process as any other proposal. The types of constructive changes may include:

- Defective plans and specifications
- Engineer's interpretation
- Higher standard of performance than specified
- Improper inspection and rejection
- Change in method of performance
- Change in the construction sequence
- Owner nondisclosure
- Impossibility/impracticability of performance

Constructive changes are a major source of construction disputes. A constructive change arises when the contractor alleges that something that the owner has done, or failed to do, has resulted in a *de facto* change in the contract requirements. The argument, of course, is that the contractor is entitled to extra compensation for performing the work. The owner frequently disagrees that a change in the contract requirements has occurred.

Most constructive changes disputes center around the interpretation of the plans and specifications. It is not surprising that owners and architect/engineers tend to interpret the contract in the manner that will be most beneficial to the project. Contractors, on the other hand, are inclined to read the plans and specifications in a manner that will minimize performance costs. Obviously, the most effective way to prevent constructive change disputes is to have a detailed, carefully prepared scope of work. This does not always happen, of course. Even when it does, disputes may still arise. It is for this reason that a comprehensive changes clause should be included in the General Conditions of any construction contract. These clauses should specify the agreed-upon change to the contract and should include the following information:

- Identification of change order
- Description of change
- Reason for change
- Change in contract price
- Change in unit prices (if applicable)
- Change to contract time
- Statement that secondary impacts are included
- Approvals by owner's and contractor's representatives

The owner may require the engineer's signature for internal control, but there is no substitute for the owner's signature. A change order is a written agreement between the owner and the contractor authorizing an addition, deletion, or revision in the work and/or time of completion within the limits of the terms of the construction contract after it has been executed. It is a specific type of contract modification that does not go beyond the general scope of the existing contract. The typical change order generally originates as a claim or recommendation for a change from the contractor or as a request-for-proposal from the owner, seeking a change to the existing contract documents.

Evaluation

A contract change order is always used to effect a change within a contract. Such change order evaluation should be used to change contract plans or to specify the method and amount of payment and changes in contract time therefrom; to change contract specifications, including changes in payment and contract time that may result from such changes; to effect agreements concerning the order of the work, including any payment or changes in contract that may result; for administrative purposes, to establish the method of extra work payment and funds for work already stipulated in the contract; for administrative purposes, to authorize an increase in extra work funds necessary to complete a previously authorized change; to cover adjustments to contract unit prices for overruns and underruns, when required by the specifications; to effect cost reduction incentive proposals (value engineering proposals); and to effect payment after settlement of claims.

A contract change order is used in most instances when a written agreement by both parties to the contract is either necessary or desirable. Such use further serves the purpose of notifying a contractor of his or her right to file a protest if he or she fails to execute a change order. The following are common categories or conditions that generally give rise to the evaluation of need for a contract change order:

- Differing site conditions
- Errors and omissions in plans and specs
- Changes instituted by regulatory agencies
- Design changes
- Overruns/underruns in quantities beyond limits
- Factors affecting time of completion

Before it can be determined that a delay in the work was compensable, thereby justifying the issuance of a change order granting both additional time and money to the contractor, the following questions need to be answered:

- Was the cause of the delay beyond the contractor's control? Did the contractor fail to take normal precautions?
- Was the contractor ready and able to work?
- Did the contractor submit a detailed schedule projecting project completion within the allotted time? Was the schedule updated regularly? Did the updated schedule justify a time extension?
- Did this schedule contain a critical path analysis or equivalent?
- Has the contractor maintained sufficient forces in those operations along the critical path where needed to meet target dates?
- How have causes, other than normal weather, beyond the control and without the fault or negligence of the contractor affected target dates along the critical path?
- Was the contractor shut down for other reasons?

Change Orders for Differing Site Conditions

Once the information on differing site conditions has been assembled, the owner should be in a position to determine if the site conditions differed materially from those indicated in the contract or those ordinarily encountered and whether this will justify an increase in the construction contractor's cost or the time to complete the work. The owner or its architect/engineer should be able to determine whether a change order is justified by a careful review of the contractor's supporting documentation, including reports, pictures, plans, and whatever is appropriate to demonstrate where and how actual conditions deviated from the plan or from conditions that should have been anticipated.

Justification for the issuance of a change order to compensate the contractor for a differing site condition will be influenced by the specific terms of the contract documents. Some guidelines follow that suggest special areas of concern to be evaluated prior to issuance of a change order to cover extra work resulting from a differing site condition:

1. What is the differing site condition?
2. What is the site condition shown on the plans?
3. What sections of the specifications are applicable?

4. Do the parties have the latest revised issues of the plans and specifications?
5. Have the parties checked all of the plan references? Plans are usually grouped by specialty: architectural, site, plumbing, mechanical, structural, electrical, special, and standard details. A critical note or dimension on one of the specialty drawings may solve the whole puzzle of need and responsibility.
6. Has the inspection staff provided adequate data? If not, correct that situation immediately. The next claim could be much larger.
7. Why was the condition shown differently by the design?
8. Was provision made for this situation in the contract documents?
9. Did the construction contractor encounter unstable soils, rock excavation, or subsurface structures where no careful prebid site inspection and contract documents could have predicted their existence?

Subsurface Investigations

Prior to the determination of the validity of a contractor's request for a change order to compensate the contractor for a differing site condition, the architect/engineer and the owner should consider the following questions before assuming that the responsibility is all on the shoulders of the contractor:

1. Were borings made or test holes dug, displayed correctly, available to all bidders, of proper kind and depth, spaced at reasonable intervals? Contract documents often have disclaimers regarding differing subsurface conditions, but these are not always as enforceable as they may appear.
2. Was the overall subsurface investigation in proportion to the type and magnitude of the project?
3. Was the geologic history of the site incorporated into the subsurface investigation data displayed in the contract documents?
4. Although not part of the evaluation, the owner should consider whether additional subsurface investigation should be initiated immediately to minimize future claims and delays.

Inclusion of Soils Reports in Contract Documents

Whenever the owner or the architect/engineer knows of or is in the possession of soils reports, logs of borings, asbestos reports, hazardous waste reports, or similar knowledge about a construction site, it is under obligation to make full disclosure of all such documents to the bidders on a project. This is normally done through the medium of identifying them by title, origin, and date in the Notice Inviting Bids, and at the same time cautioning the bidder that they are not a part of the contract documents and are identified for the convenience of the bidders. Usually, exculpatory language is included to the effect that the owner or its architect/engineer makes no representations regarding such data, and that the bidder must interpret and draw his or her own conclusions therefrom. If the contract documents state that the soils reports, for example, should not be a part of the contract documents, inclusion of such documents as a part of the contract documents recordkeeping invariably lead to claims for extra work and change orders.

Other Influences on Change Order Justification

Prior to filing a request for a change order due to alleged differing site conditions, the architect/engineer should determine whether or not the contractor complied with the following:

1. Did the construction contractor promptly and before such conditions were disturbed comply with the owner's modification requirements?
2. Did the construction contractor make a site inspection and evaluation of conditions prior to bidding?
3. Is the construction layout correct? Who performed the construction staking; who checked the staking?
4. Did utility companies respond and locate their respective facilities? Did the construction contractor protect or obliterate utility reference stakes and use suitable construction techniques to protect structures and utilities?

Initiation of Change Orders

Change orders are usually initiated by construction personnel at the project site. However, changes are also requested from various other sources, such as the contractor, the design firm, outside public agencies, or private individuals. In any case, they must be authorized by the owner before any change order authorizing extra work is valid. A proposed contract change order is written only after the designers have given consideration as to the necessity, propriety, other methods of accomplishing the work, method of compensation, effect on contract time, estimate of cost, the contractor's reaction to the proposed change, and the probability of final approval.

Any change in the work that involves a change in the original contract price must be approved in writing by the owner before a change order can be executed. If it is not the owner who signs, then the party who does sign for him or her must have written authorization from the owner to sign on his or her behalf. The architect or engineer of record, by virtue of their positions alone, have no authority to order changes to the contract. If they do act in this manner, they must be authorized in some way to act in the owner's behalf.

The change order must be clear, concise, and explicit. It must tell the contractor what is to be done, where or within what limits, when the work is to be performed, if the order of the work is affected, how the contractor will be paid, and what consideration will be given to contract time (extensions, etc.). For users of the current EJCDC documents, the Work Change Directive was developed for use in situations involving changes in the work which, if not processed expeditiously, might delay the project. The AIA counterpart is called a Construction Change Directive. Such changes are often initiated in the field and may affect contract price or time. This form is not a change order, however, but only a directive to proceed with the work that may be included in a subsequent change order. Unfortunately, the AIA form confuses the issue somewhat by having a block where the contractor can sign the document. If both parties are to sign the change document, you may as well issue a regular change order; after all, the purpose of that document is to allow the issuance of a unilateral order to perform the work. The EJCDC, AIA, and Federal forms for implementing changes in the contract are:

Bilateral Contract Modification

- Change Order EJCDC Document 1910-8-B
- Change Order AIA Document G701
- Federal Contract Modification
- Change Order-Fisk Form 15-2

Unilateral Contract Modification

- Work Change Directive EJCDC Document 1910-8-F
- Construction Change Directive AIA Document G714
- Federal Change Order
- Work Directive Change-Fisk Form 15-3

Field Orders

Although the use of the change order and the Work Change Directive is quite clear and it satisfies all the requirements for documenting and executing an amendment to the contract provisions, the same cannot necessarily be said of a Field Order. The Field Order is supposed to be an order from the architect/engineer to the contractor that either interprets the contract documents or orders minor changes in the work without changing either the contract sum or the contract time. As a document to interpret the contract provisions, it serves its purpose admirably well. However, any document that changes any of the provisions of the plans, specifications, or other contract documents is a change order no matter what you may choose to call it. As such, in the absence of a written authorization from the owner that allows the architect/engineer unilaterally to make such changes, only the owner can execute this document.

On a recent project where this document was used, it was desired to make a minor change in the work, for which there would be no additional charge by the contractor. When confronted with this document, the contractor asked that a formal change order be issued instead, even though the terms had been agreed to previously and no charge was being made for the changes. The contractor was not only legally right in requesting a formal change order, but wise to do so. A formal change order provides the written authorization of the owner to make a change in the contract provisions; it states the effect on the contract sum, if any; it specifies the effect on contract time, if any; and it could serve to release the contractor from liability that might result from making the change.

Any document that alters the terms of the original contract in any way must be considered as a change order. As such, the document must either be signed by the owner or by someone with the power of attorney to act for him. The architect/engineer's Field Order only provides for the signature of the architect/engineer and thus does not satisfy this requirement unless the architect/engineer also possesses written authority of the owner to execute such changes on his or her behalf. Even then, if a change in the work is intended, as opposed to an interpretation of the contract documents, it is still far more desirable to utilize the change order form for this purpose as a protection to all parties to the contract, even though there will be no change in contract price or time. The AIA Construction Change Directive differs, principally, from the form of the EJCDC Work Change Directive in that it provides for the written acceptance by the contractor. There are only four basic steps involved in the execution of a nonemergency change order, although the in-house procedures of an organization may expand on this to satisfy their own procedural requirements. The four basic steps are:

1. Change order request by initiator (may be initiated by contractor or architect/engineer) for architect/engineer and owner's approval.
2. Upon approval of initiator request, discuss with contractor and draft a change order proposal document citing the effect of the change on contract time and cost.
3. Contractor submits a signed change order proposal to the owner, showing all costs and additional time required.
4. Owner accepts proposal by signing the proposal and orders execution of the work described.

Cost of Delays

Often, the execution of a change order involves slowdowns or delays of the contractor's operations. It should be kept in mind that all such delays involve extra cost to the contractor and, as such, the costs are recoverable by the contractor. This lends even more importance to the contractor's obligation to record all equipment and workers at the site, and to note whether such equipment is being used on that day or not. Such information can be the basis for defending claims of excessive charges for delays caused by extra work.

Construction contracts are subject to broader principles of interpretation than most other contracts. There are five basic rules of contract interpretation that are quite important in construction contracts. The courts are frequently unfamiliar with the specialized rules that have evolved in the construction industry and often rely upon the testimony of experts in the subject area to guide them in forming a decision. A written contract is merely a form of documentation of the conditions agreed upon by the parties. However, it reflects the understanding that each party had with regard to the wording of the documents. If every written contract were clear, unambiguous, and complete, the rules of contract interpretation would be unnecessary, as the intent and understanding of the parties would be self-evident. Unfortunately, this is not the case.

In any construction contract, it is almost inevitable that the written documents will not adequately address every single matter. There may be gaps, conflicts, or subtle ambiguities. The five principles mentioned here are used to help resolve such problems. They are little more than applications of common sense in an effort to determine the intentions of the parties from a written document that does not adequately express those intentions. Often in a construction dispute, each party concentrates on some narrow provision of the specifications, general conditions, or drawings that seems to support his or her position. The attitude is: "There it is in writing." The law recognizes that the intentions of the parties cannot be determined by examining one small part of a document by itself. Every provision of the document is presumed to have meaning; therefore, no provision should be presumed to be meaningless or superfluous. The document must be considered in its entirety.

There is a well-known principle that the document will be construed against the drafter, and thus must be recognized in any dispute as to the meaning of the wording used in the document. There may be latent ambiguity; the contract may be subject to more than one reasonable interpretation, and a court is often faced with the burden of choosing between the two. In such a situation, the document is interpreted in favor of the party that did not draft the document. The reasoning behind this principle is that the party that drafted the document had ample opportunity to avoid ambiguity and clearly express the intended meaning, but failed to do so. The nondrafting party did not have such an opportunity; they reasonably assumed that the contract language meant a certain thing, and in court that interpretation must prevail.

This means that in construction contracts the contractor's reasonable interpretation will usually prevail over that of the owner or the architect/engineer, and in the case of prime contractor/subcontractor relations, the subcontractor's reasonable interpretation will usually prevail over that of the prime contractor. Although not necessarily so in all cases, contracts are usually prepared by owners or architect/engineers, and subcontracts are usually prepared by prime contractors.

"The document supersedes all previous discussions" implies that any oral commitments made prior to the execution of an agreement are nonbinding, as the written document is presumed to constitute the entire agreement. Basically, the rule is that the document speaks for itself. However, when it does not speak for itself, or when it is so inarticulate that no one knows what it means, oral testimony is permitted to determine the intended meaning of the parties. This is a well-known exception to the payroll evidence rule. After hearing the expert testimony, the final interpretation of the contract will probably be decided based upon the respective credibility of each party's expert testimony. Unfortunately, this does not assure that justice will be served.

Specific Terms Govern over General Terms

Sometimes there will be a conflict between two different provisions of a written agreement. The court has ruled that the more specific term should govern over the general term. The reasoning behind this is that a narrowly drafted provision was customized to fit a specific situation and thus reflects more accurately the parties intentions than a "general" or "boilerplate" provision of the contract. On construction contracts, this is often applied by making handwritten or typewritten provisions prevail over conflicting provisions found in the preprinted sections of the contract. Similarly, job-specific Supplementary General Conditions will prevail over the General Conditions.

Documents must be read in the context of the trade. This is another common-sense rule. If a word or phrase has a commonly accepted meaning in the construction industry, that meaning must be applied when reading and interpreting construction contract documents.

Construction Problems

Every construction project seems to include its share of problems. In one way of thinking, perhaps if it were not for these problems, the job of prime contractor might never have been conceived. A good prime contractor is not simply a person who has the unique ability to solve problems, for generally the subcontractor can do this. It is the prime contractor's ability to apply personal experience and knowledge to the project, to look ahead, anticipate the occurrence of various events resulting from the various approaches that the contractor might take, evaluate these conditions, and offer constructive assistance in seeing that the potential problems either never occur or are minimized. To understand fully a prime contractor's application to the terms of the construction contract, several terms frequently used in this book will be defined. This section is divided into two major classifications. The first deals with the definition and the administrative procedures involved in the handling of claims under the terms of the construction contract; the second addresses itself to the methods available for the settlement of such differences, as well as an explanation of the practical pros and cons of arbitration under the rules of the American Arbitration Association, which is receiving growing respect as a fast, economical, and fair method for the resolution of contract disputes.

Protests

The term protests as used here refers to disputes arising out of the issuance of a contract change order by the architect or engineer against the objections of the contractor. The terms of the contract documents for the orderly filing of written protests by the contractor as a means of establishing its claims for additional compensation or time to complete the work or claims of unfair treatment under the terms of the contract, short of turning the problem over to arbitration, is normally covered in the General Conditions (boilerplate). In Articles 9.11 through 9.12 and Article 16 (including Exhibit GC-A) of the EJCDC Standard General Conditions of the Construction Contract, and in Sub-Clauses 67.1 through 67.4 of the EJCDC Conditions of Contract, the procedures for the handling of protests and disagreements are specified in detail, up to and including referring the matter to binding arbitration. (See Appendix for detailed EJCDC contract documents classifications breakout.)

The normal procedure is for the prime contractor (GC) to discuss with the owner any objections to a particular change order or other directions of the architect/engineer. If the contractor's objections can be satisfied by minor changes in the provisions of a proposed change order or direction and such changes do not violate the contract provisions (this includes plans, specifications, addenda, previous change orders, codes, permit provisions, or similar constraints), the change should normally be discussed with the architect/engineer or owner and with their concurrence the change should normally be made. This will often avoid a formal protest, which is often costly even to the winner. In the event that the owner cannot satisfy the contractor's objections, the contractor should explore other possible means of settlement. When neither the architect/engineer or owner can resolve the problem through normal procedures, the pending change order or other direction of the architect/engineer or the owner should be issued in its original form and the contractor reminded of the provisions of the contract documents that require the contractor's conformance with any such order of the architect/engineer or owner, even if the contractor disagrees with it, under penalty of being considered in default on the contract.

It should also be carefully noted in contractor's job diary that such a reminder was given to the contractor under protest. This does not take away the contractor's right to file a letter of protest that could be used as a later basis of claims against the owner, provided that such letter is submitted within the time schedules that are specified in the General Conditions of the construction contract.

When a protest letter is received from the contractor, it will be examined carefully prior to acknowledgment of its receipt to assure that the basic requirements of the specifications are included in the contractor's letter. The architect/ engineer or owner will then review the merits of the contractor's protest. If the architect/engineer or owner decides that the protest is without merit, they should issue a letter to the contractor advising of its rights under the contract to file legal claims. Normally, if the contractor fails to file claims in accordance with the provisions of the contract, the contractor is in default or breach of contract. The letter of protest should follow the dispute resolution guidelines set forth in the example below of the industry standard EJCDC No. 1910-8 (1990 ed.): Provisions for dispute resolution in the 1990 edition of the EJCDC General Conditions of the Construction Contract.

OWNER and CONTRACTOR hereby agree that Article 16 of the General Conditions to the Agreement between OWNER and CONTRACTOR is amended to include the following agreement of the parties:

16.1. All claims, disputes and other matters in question between OWNER and CONTRACTOR arising out of or relating to the Contract Documents or the breach thereof (except for claims which have been waived by the making or acceptance of final payment as provided by paragraph 14.1.5) will be decided by arbitration in accordance with the Construction Industry Arbitration Rules of the American Arbitration Association then obtaining, subject to the limitations of this Article 16. This agreement so to arbitrate and any other agreement or consent to arbitrate entered into in accordance herewith as provided in this Article 16 will be specifically enforceable under the prevailing law of any court having jurisdiction.

16.2. No demand for arbitration of any claim, dispute or other matter that is required to be referred to ARCHITECT initially for decision in accordance with paragraph 9.11 will be made until the earlier of (a) the date on which ARCHITECT has rendered a written decision or (b) the thirty-first day after the parties have presented their evidence to ARCHITECT if a written decision has not been rendered by ARCHITECT before that date. No demand for arbitration of any such claim, dispute or other matter will be made later than thirty days after the date on which ARCHITECT has rendered a written decision in respect thereof in accordance with paragraph 9.1.1; and the failure to demand arbitration within said thirty days' period will result in Owner/Architect's decision being final and binding upon OWNER and CONTRACTOR. If ARCHITECT renders a decision after arbitration proceedings have been initiated, such decision may be entered as evidence but will not supersede the arbitration proceedings, except where the decision is acceptable to the parties concerned. No demand for arbitration of any written decision of ARCHITECT rendered in accordance with paragraph 9.10 will be made later than ten days after the party making such demand has delivered written notice of intention to appeal as provided in paragraph 9.10.

16.3. Notice of the demand for arbitration will be filed in writing with the other party to the Agreement and with the American Arbitration Association, and a copy will be sent to OWNER for information. The demand for arbitration will be made within the thirty-day or ten-day period specified in paragraph 16.2 as applicable, and in all other cases within a reasonable time after the claim, dispute or other matter in question has arisen. and in no event shall any such demand be made after the date when institution of legal or equitable proceedings based on such claim, dispute or other matter in question would be barred by the applicable statute of limitations.

16.4. Except as provided in paragraph 16.5 below, no arbitration arising out of or relating to the Contract Documents shall include by consolidation, joiner or in any other manner any other person or entity (including OWNER, Owner's Consultant and the officers, directors, agents, employees or consultants of any of them) who is not a party to this contract unless:

16.4.1. The inclusion of such other person or entity is necessary if complete relief is to be afforded among those who are already parties to the arbitration, and

16.4.2. Such other person or entity is substantially involved in a question of law or fact which is common to those who are already parties to the arbitration and which will arise in such proceedings, and

16.4.3. The written consent of the other person or entity sought to be included and of OWNER and CONTRACTOR has been obtained for such inclusion, which consent shall make specific reference to this paragraph; but no such consent shall constitute consent to arbitration of any dispute not specifically described in such consent or to arbitration with any party not specifically identified in such consent.

16.5. Notwithstanding paragraph 16.4 if a claim, dispute or other matter in question between OWNER and CONTRACTOR involves the Work of a Subcontractor, either OWNER or CONTRACTOR may join such Subcontractor as a party to the arbitration between OWNER and CONTRACTOR hereunder. CONTRACTOR shall include in all subcontracts required by paragraph 6.11 a specific provision whereby the Subcontractor consents to being joined in an arbitration between OWNER and CONTRACTOR involving the Work of such Subcontractor. Nothing in this paragraph 16.5 nor in the provision of such subcontract consenting to joiner shall create any claim, right or cause of action in favor of Subcontractor and against OWNER, ARCHITECT, or Owner's Consultants that does not otherwise exist.

16.6. The award rendered by the arbitrators will be final, judgment may be entered upon it in any court having jurisdiction thereof, and it will not be subject to modification or appeal.

16.6.1 OWNER and CONTRACTOR agree that they shall settle all unsettled claims, counterclaims, matters in question between them arising out of Contract Documents or the breach thereof by The American Arbitration As Construction Industry Mediation Rules of the American Arbitration Association prior to either of them initiating against the other a demand for arbitration pursuant to paragraphs 16.1 through 16.6, unless delay in initiating arbitration would irrevocably prejudice one of the parties. The respective thirty and ten day time limits within which to file a demand for arbitration as provided in paragraphs 16.2 and 16.3 above shall be suspended with respect to a dispute submitted to mediation within those same applicable time limits and shall remain suspended until ten days after the termination of the mediation. The mediator of any dispute submitted to mediation under this Agreement shall not serve as arbitrator of such dispute unless otherwise agreed.

17. 1. If and to the extent that OWNER and CONTRACTOR have agreed on the method and procedure for resolving disputes between them that may arise under this Agreement, such dispute resolution method and procedure, if any, shall be as set forth in Exhibit GC-A, "Dispute Resolution Agreement," to be attached hereto and made a part hereof. If no such agreement on the method and procedure for resolving such disputes has been reached, and subject to the provisions of paragraphs 9.10, 9.11, and 9.12, OWNER and CONTRACTOR may exercise such rights or remedies as either may otherwise have under the Contract Documents or by laws or regulations in respect of any dispute.

Decisions on Disputes

9.11. OWNER will be the initial interpreter of the requirements of the Contract Documents and judge of the acceptability of the Work thereunder, Claims, disputes and other matters relating to the acceptability of the Work or the interpretation of the requirements of the Contract Documents pertaining to the performance and furnishing of the Work and Claims under Articles 11 and 12 in respect of changes in the Contract Price or Contract Times will be referred initially to OWNER in writing accompanying a request for a final decision in accordance with this paragraph.

Written notice of each such claim, dispute or other matter will be delivered by the claimant to OWNER/ARCHITECT and the other party to the Agreement promptly (but in no event later than thirty days) after the start of the occurrence or event giving rise thereto, and written supporting data will be submitted to OWNER/ARCHITECT and the other party within sixty days after such occurrence or event unless OWNER/ARCHITECT allows an additional period of time for the submission of additional or more accurate data in support of such claim, dispute or other matter.

The opposing party shall submit any response to OWNER/ARCHITECT and the claimant within thirty days after receipt of the claimant's last submittal (unless OWNER/ARCHITECT allows additional time). OWNER/ARCHITECT receives a formal decision in writing within thirty days after receipt of the opposing party's submittal, in accordance with this paragraph.

Potential Claims

The term potential claim applies to any differences arising out of the performance of the work that might reasonably lead to the later filing of a formal claim by the contractor if the difference cannot be resolved in the field.

It should be the policy of the architect/engineer or owner to consider the merits of a potential claim at the earliest possible time. As soon as the contractor has knowledge of the existence of a dispute that may lead to the filing of a potential claim, the situation should be discussed with the owner. If the owner determines that the contractor's preliminary arguments are valid, such corrective measures should be taken as are within the scope of the owner's authority under the contract, including the possibility of making recommendation to the architect/engineer owner to submit change orders to alleviate the problem. In the event that the contractor cannot resolve the differences in the field, the problem should be discussed with the architect/engineer or the owner. If the differences still cannot be resolved within the terms of the contract, the contractor should make use of the provisions of the contract documents relating to the time and methods for it to file claims.

Upon receipt of a potential claim in writing from the contractor, and prior to submitting it to the architect/engineer for review, it should be reviewed by the owner to see that it contains the basic information necessary, such as the reasons that the contractor believes additional consideration is due. It should also be reviewed to determine that the timeliness of the submittal is in accordance with the terms of the contract documents. If the contractor's letter is not sufficiently complete with respect to the nature of the claim, the architect/engineer or owner should request in writing that the contractor submit additional data before further processing of the contractor's claim.

Each potential claim should be reviewed by the architect/engineer or owner. If after this review the potential claim is considered as being without merit, an answering letter should be sent to the contractor. The letter should include the statements referred to under "protests" in previous paragraphs. If the contractor's notice of potential claim does not meet the requirements for timeliness as set forth in the specifications, the architect/engineer or owner should advise the contractor in writing of this deficiency.

Early Claims Reporting

The professional liability underwriters would prefer to have claims overreported rather than underreported, as it is better to report a dispute situation before it becomes an actual claim than to wait until it is a very hot issue and the battle lines have been drawn. The claims department of a professional liability insurance company can usually give counsel on methods of reducing exposure, or at least on keeping things fair. In addition, their claims department can advise on methods for mitigating damages (if any) so that whoever is finally held responsible, the financial burden will be less. By spotting a dispute before it becomes a claim and reporting it to the architect/engineer or owner, a monumental amount of later grief may be avoided.

Insurance Claims

Sometimes, through oversight or lack of understanding of contract practice, issues can arise that may spell latent financial disaster for the owner. On one occasion, an insurance company in southern California investigated a project that had just suffered major fire damage resulting from an arsonist who set the blaze about 4:00 p.m. on a Thanksgiving day when no one was at the site, resulting in a near total loss of a project that was at that time about 70% completed. The initial thrust of the investigation was to determine whether the project was a total loss, hence a higher payoff by the insurance carrier, or whether any remaining portions of the structure were salvageable. Copies of the policies and project drawings were made available and a detailed field investigation was started. The police crime laboratory made a thorough examination of the structural properties of the remaining portions of the concrete and concrete block portions of the structures and an insurance investigation was made with a determination that there was a salvage value of over $300,000.

All wood and steel members were a total loss, and the only surviving remnants of the structure were the reinforced concrete parking garage structure and some peripheral concrete block buildings. The surprise came, however, when the insurance policy and the construction contract documents were examined. The policy involved appeared to be a standard builder's-risk fire policy. Although it was purchased by the owner, not the contractor, this is not unusual. Often, the owner buys the policy; otherwise, the contractor buys the policy. The problem occurred because the policy terms and conditions were not included in the construction contract documents by the architect, and the endorsements called for a 6-ft chain-link fence to be erected around the project and 24-hour security guard service.

Investigation showed that the project was fenced but no security guard service was provided. However, because the terms of the policy were not specified as a part of the contractor's construction documents, it was not the contractor's responsibility to provide either guard service or fencing. The contractor provided fencing on its own to protect its equipment, but was possibly completely unaware of the insurance endorsement requirements. Thus, the obligation to assure that the job was properly policed rested completely upon the shoulders of the owner, who unknowingly thought the contractor was obligated to provide security guards. The claims adjustment recommendation here was that the insurance company deny the entire claim because of the breach. A court of law later reached the same conclusion.

The lesson here is that whenever the owner elects to purchase its own policy of insurance on a construction project, it should make certain that it includes the terms and conditions imposed in the policy as a part of the contract obligation under the construction contract.

Home Office Overhead

The costs of delays involve many elements. Naturally, the direct costs are affected, but a contractor's work efficiency, construction schedule, available favorable weather, impact costs, and even home office overhead may be affected. Normally, most contract documents specify the limits of overhead charges that may be levied for force account work, but it should be kept in mind that such limitations apply only to a contract that is completed within the specified time. If the owner causes a compensable delay that forces the contract into a longer time frame, the contractor may be entitled to unabsorbed or extended home office overhead.

Recovering damages for home office overhead requires proof of actual damage. Although the cost of home office overhead is well recognized in a suspension-of-work situation, the contractor must still show that it was unable to fill in with other work during the suspension period.

of the following, on every job, until you dump your records on a project (normally seven years). It's Murphy's law: If you can contact these people for depositions or subpoena, you won't need to. If you haven't a got a clue as to who they are or where they are now, that's when you'll need them the most.

- Contractor's project personnel
- Owner's project personnel
- Architect/engineer personnel
- Construction management personnel
- Subcontractors
- Inspectors
- Suppliers
- Testing lab personnel
- Consultants
- Expert witnesses
- Additional site-related witnesses
  (Licensed or certified helps)

Appendix

A

EJCDC Table of Code Contract Standards

The following appendix contains the table of standards set for general conditions of the construction contract that is prepared by the Engineers Joint Contract Documents Committee, recognized nationwide as the authoritative industry standard for construction contracts. The enforcement of your contract documents validity in construction claims litigation or arbitration lies within the parameters of these accepted industry standards. Use this table to locate the contract number and legal precedence of the item you seek.

This table of the Standard General Conditions of The Construction Contract, prepared by the EJCDC, is current as of the time of this writing.

Article #	Division	Title	Page
1.1	Contract Definitions	Addenda	13
1.2	Contract Definitions	Agreement	13
1.3	Contract Definitions	Payment	13
1.4	Contract Definitions	Asbestos	13
1.5	Contract Definitions	Bid	13
1.6	Contract Definitions	Bid Docs	13
1.7	Contract Definitions	Bid Required	13
1.8	Contract Definitions	Bonds	13
1.9	Contract Definitions	Change Order	13
1.10	Contract Definitions	Contract Docs	13
1.11	Contract Definitions	Contract Price	13
1.12	Contract Definitions	Contract Times	13
1.13	Contract Definitions	Contractor	13

1.14	Contract Definitions	Defective	13
1.15	Contract Definitions	Drawings	13
1.16	Contract Definitions	Agreement Date	13
1.17	Contract Definitions	Engineer	13
1.18	Contract Definitions	Consultant	13
1.19	Contract Definitions	Field Order	13
1.20	Contract Definitions	General	14
1.21	Contract Definitions	Hazardous	14
1.22	Contract Definitions	Laws/Regulations	14
1.23	Contract Definitions	Liens	14
1.24	Contract Definitions	Milestones	14
1.25	Contract Definitions	Award Notice	14
1.26	Contract Definitions	Proceed Notice	14
1.27	Contract Definitions	Owner	14
1.28	Contract Definitions	Partial Utilization	14
1.29	Contract Definitions	PCBs	14
1.30	Contract Definitions	Petroleum	14
1.31	Contract Definitions	Project	14
1.32	Contract Definitions	Radioactive	14
1.33	Contract Definitions	Representative	14
1.34	Contract Definitions	Samples	14
1.35	Contract Definitions	Shop Drawings	14
1.36	Contract Definitions	Specifications	14
1.37	Contract Definitions	Subcontractor	14
1.38	Contract Definitions	Substantial Comp.	14
1.39	Contract Definitions	Supplementary	14
1.40	Contract Definitions	Supplier	14
1.41	Contract Definitions	Underground	14
1.42	Contract Definitions	Unit Price Work	14
1.43	Contract Definitions	Work	15
1.44	Contract Definitions	Work Change	15
1.45	Contract Definitions	Written Amend.	15
2.0	Preliminary Matters	Delivery of Bonds	15
2.1	Preliminary Matters	Copies Documents	15
2.2	Preliminary Matters	Commencement	15
2.4	Preliminary Matters	Starting Work	15
2.5	Preliminary Matters	Schedule	15

2.7	Preliminary Matters	PreCon	15
2.9	Preliminary Matters	Accepted Schedule	16
3.0	Contract Documents	Standards	16
3.1 - 3.2	Contract Documents	Intent	16
3.3	Contract Documents	Specifications	16
3.4	Contract Documents	Terms/Adjectives	17
3.5	Contract Documents	Amending	17
3.6	Contract Documents	Supplementing	17
3.7	Contract Documents	Reuse Docs.	17
4.0	Physical Conditions	Reference Points	17
4.1	Physical Conditions	Land Availability	17
4.2	Physical Conditions	Subsurface	17
4.2.1	Physical Conditions	Reports/Drawings	17
4.2.2	Physical Conditions	Reliance	18
4.2.3	Physical Conditions	Differing Conditions	18
4.2.4	Physical Conditions	Engineer's Review	18
4.2.5	Physical Conditions	Contract Change	18
4.2.6	Physical Conditions	Price Change	18
4.2.8	Physical Conditions	Time Adjustments	18
4.3	Physical Conditions	Underground	18
4.3.1	Physical Conditions	Shown/Indicated	18
4.3.2	Physical Conditions	Not Shown	19
4.4	Physical Conditions	Reference Points	19
4.5	Physical Conditions	Hazardous Materials	19
5.0	Bonds & Insurance	Bonds	20
5.1	Bonds & Insurance	Payment Bonds	20
5.2	Bonds & Insurance	Performance Bond	20
5.3	Bonds & Insurance	Sureties	20
5.4	Bonds & Insurance	Contractor's Ins.	20
5.5	Bonds & Insurance	Owner's Ins.	21
5.6	Bonds & Insurance	Property Ins.	21
5.7	Bonds & Insurance	Machinery	21
5.8	Bonds & Insurance	Cancellation	21
5.9	Bonds & Insurance	Contractor	22
5.10	Bonds & Insurance	Special Ins.	22

15.1	Work Suspension	Owner Suspension	40
15.2 - 15.4	Work Suspension	Termination	40
15.5	Work Suspension	Contractor	41
16.0 - 16.9	Dispute Resolution	Precedence	41
17.0	Miscellaneous	Extras	42
17.1	Miscellaneous	Giving Notice	42
17.3	Miscellaneous	Time Computation	42
17.4	Miscellaneous	Notice of Claim	42
17.5	Miscellaneous	Remedies	42
17.6	Miscellaneous	Professional Fees	42
17.18	Miscellaneous	Court Costs	42
18.0	Acceptance	Bonds/Insurance	43
18.1	Acceptance	Defective Work	43
18.2	Acceptance	Final Payment	43
18.3	Acceptance	Insurance	43
18.4	Acceptance	Other Work	43
18.5	Acceptance	Substitutes	43
18.6	Acceptance	Work By Owner	44
19.0	Access	Lands	44
19.1	Access	Site	44
19.3	Access	Work	44
20.0	Acts/Omissions	Contractor	45
20.1	Acts/Omissions	Engineer	45
20.2	Acts/Omissions	Architect	45
20.3	Acts/Omissions	Owner	45
20.4	Acts/Omissions	Inspector	45
21.0	Addenda	Definition	46
21.1	Addenda	Specifications	46
21.2	Addenda	Insurance	46
21.3	Addenda	Adjustments	46
21.5	Addenda	Prices	46
21.6	Addenda	Contract Times	46

21.7	Addenda	Adjustments	46
21.8	Addenda	Progress	47
21.9	Addenda	Agreement	47
22.0	Amendment	Written	47
22.1	Amendment	General	47
22.2	Amendment	Claims	47
22.3	Amendment	Contractor	47
22.4	Amendment	Owner	47
22.5	Amendment	Architect	48
22.6	Amendment	Engineer	48
22.8	Amendment	Definition	48
22.9	Amendment	Prices	48
23.0	Authorization	Owner	48
23.1	Authorization	Architect	48
23.2	Authorization	Engineer	49
23.3	Authorization	Contractor	49
23.4	Authorization	Prior	49
23.5	Authorization	During Const.	49
23.6	Authorization	Change Order	49
23.7	Authorization	Claims	49
24.0	Contract Specs	Documents	50
24.1	Contract Specs	Bidding	50
24.2	Contract Specs	Award	50
24.3	Contract Specs	Recordation	50
24.4	Contract Specs	Bonds/Insurance	50
24.5	Contract Specs	Amendment	50
24.6	Contract Specs	Arbitration	50
24.7	Contract Specs	Changes	51
24.8	Contract Specs	Change Order	51
24.9	Contract Specs	Claims	51
25.0	Certificates	Bonds	51
25.1	Certificates	Insurance	51
25.2	Certificates	Cancellation	51
25.3	Certificates	Cash Allowance	52

71.8	Progress	Weather	89
71.9	Progress	Completion	89

Appendix

B

Contract Abbreviations

Construction contract standards are based on terminology governed by the American Institute of Architects (AIA) contract document specifications. These AIA acronyms and abbreviations are typical throughout the industry and are commonly used in residential construction contracts. The astute contractor may not know everything about contracts but he or she does know where to look it up. This is one such *get-it-right-the-first-time* reference appendix. This is supplied as an added reader bonus on suggestion of my senior editor. She thought the reader should get as many solutions to modern claims potentials as we could pack into the book. It was a good call in this builder's opinion, so this added appendix represents another key for the lock. At no extra charge I might add, Revered Reader, compliments of my editor. (However, if you write to her in thanks, don't mention my name. She may go for her gun. . .)

You will note some of these abbreviations have periods while some don't. This is not a typo, this is the way they appear in use. Don't ask me why—I don't know why. Neither does anyone else at the AIA headquarters office. It's just the way it is. Probably one of those things that this industry's tradition has passed on to us just to drive people nuts when they try to learn our system. If you're a contractor with some history under your belt, you know exactly what I mean. You will see many of these in production plans and contract documents. Always make sure you fully understand all acronyms and abbreviations that are involved with your contractual obligations. You will note by examining these that there are subtleties that go with the periods, such as PCC stands for "point of compound curve" whereas P.C.C. stands for "Portland Cement Concrete." Pay attention to such details in looking at abbreviations, especially in your long-lead items.

A	Area
AASHO	American Assn. of State Highway Officials
AB	Anchor Bolt
A.B.	Aggregate Base

ABS	Acrylonitrile Butadiene Styrene (plastic sewer pipe)
AC	Alternating Current
A.C.	Asphalt Concrete
ACI	American Concrete Institute (concrete specs for concrete production)
ACP	Asbestos Cement Pipe
ACST	Acoustic
ACST PLAS	Acoustical Plastic
ACU	Air Conditioning Unit
AD	Area Drain
ADD	Addition
ADJ	Adjustable
AFF	Above Finish Floor (elevation)
AGG.	Aggregate
AH	Air Handling Unit
AIA	American Institute of Architects (contracts and design specs)
AISC	American Institute of Steel Construction (structural steel specs)
ALT	Alternate
AMT	Amount
ANGL	Angle
ANGL DF	Angle of Deflection (intersecting angle)
APP	Attactic prolypproplyene
APPX	Approximate
APT	Apartment
ARCH	Architect(s) or Architectural
ASA	American Standards Association
ASHRAE	American Society of Heating, Refrigerating and Air Conditioning Engineers
ASME	American Society of Mechanical Engineers
ASPH	Asphalt
ASSEM	Assemble
ASSOC	Associate
ASTM	American Society of Testing Materials
AT	Acoustical Tile
AUTO	Automatic

AVG	Average
AWPA	American Wood Products Association (wood standards)
AWS	American Welding Society (steel specs standards)
AWWA	American Water Work Association (water standards)
Ba	Bay
BB	Bond Beam
BC	Begin Horizontal Curve
BF	Bottom of Footing
BHP	Boiler Horsepower
BITUM	Bituminous
BLDG	Building
BLKG	Blocking
BM	Bench Mark
BOT	Bottom
B.O.W.	Back Of Walk
BP	Blueprint (refer to plans)
BRG	Bearing (load bearing)
BRG PL	Bearing Plate
BRK	Brick
BRKT	Bracket(s)
BTR	Better
>BTR	Better Than
BTU	British Thermal Unit
BU	Bureau of Standards
BVC	Begin Vertical Curve
C	Thermal Conductive
C	One Hundred
CATV	Cable Television
CB	Catch Basin
CC	Center to Center
CEM	Cement
CF	Cubic Foot
CFM	Cubic Feet Per Minute

C&G	Curb & Gutter
CC	Center to Center
CL	Center Line
CLR.	Clear
CMU	Concrete Masonry Units (masonry blocks)
CMP	Corrugated Metal Pipe
CND	Conduit
CO	Company
CONC	Concrete
CONF	Conform
CONST.	Construct
CONT.	Contours or Continuous
CONTR	Contractor
CP	Concrete Pipe
CPM	Critical Path Management
CPM	Cycles Per Minute
CS	Cast Steel
CS	Commercial Standards (as in "up to")
CSF	100 Square Feet
CSI	Construction Specifications Institute (construction specs)
CTR	Contract (as in "per")
CU FT	Cubic Foot
CU IN	Cubic Inch
CU YD	Cubic Yard
D	Drain
D	Depth
DC	Direct Current
DEG	Degree
DEPT	Department
DET.	Detail
DF	Douglas Fir (construction grade lumber)
D.I.	Drain Inlet
DIA.	Diameter
DIAG	Diagram
DIP	Ductile Iron Pipe
DISC	Disconnect

DIV	Division
DL	Dead Load
DN	Down
DP	Duplicate
DPG	Damproofing
DSB	Double Strength B Grade Glass
D4S	Dressed & Matched on 4 Sides
DU	Double Strength
DWG.	Drawing D/W Driveway
DWV	Drain, Waste, Vent Piping
DX	Duplex
E	East
EA.	Each
E to E	End to End
EC	End Horizontal Curve
EL, ELEV.	Elevation
EJ	Expansion Joint
ENCL	Enclosure
ELEC	Electric
ELVR	Elevator
EMER	Emergency
EMT	Electrical Metallic Tubing (conduit)
ENGR	Engineer
ENT	Entrance
EP	Electrical Panel
EQ	Equal
EQUIP	Equipment
EST	Estimate
ETE	End To End
EWC	Electric Water Cooling
EXC	Excavate
EXCL	Exclude
EXH	Exhaust
EXIST	Existing
EXP	Expansion
EXR JT	Expansion Joint
EXP	Exposure

EXPO	Exposed
EXST	Existing
EXT	Exterior
EXTN	Extension
E.P.	Edge of Pavement
EQ	Equal
EVC	End Vertical Curve
EX.,EXIST	Existing
F	Fahrenheit
FA	Fire Alarm
F&I	Furnish & Install
FAB	Fabricate
FAO	Finish All Over
FB	Flat Bar
FBM	Foot Board Measure
FD	Floor Drain
FD	Found
FDN	Foundation (East Coast)
FE	Fire Extinguisher
FEC	Fire Extinguisher Cabinet
FED	Federal
FED SPEC	Federal Specification
FG	Finish Grade
FH	Fire Hose
FHC	Fire Hose Cabinet
FICA	Social Security Tax
FIG	Figure
F.L.	Flowline
FIN	Finish
FIN CEIL	Finish Ceiling (elevation)
FIN FL	Finish Floor (elevation)
FIX	Fixture(s)
FL	Flashing
FL	Floor
FLASH	Roof Flashing
FLG	Flooring (type)

FLNG	Flange
FLUOR	Fluorescent
FND	Foundation (West Coast)
FOB	Free On Board
FOC	Face Of Concrete
FOF	Face Of Finish
FOS	Face Of Studs
FP	Fireplace
FPM	Feet Per Minute
FPRF	Fireproof
FPS	Feet Per Second
FR	Frame
FRG	Furring
FRP	Fiberglass Reinforced Plastic
FS	Full Size
FT	Foot
FT3	Cubic Feet
FTG	Footing
FTG	Fitting
FUT	Future (use)
FUTA	Federal Unemployment Tax
FWH	Frostproof
FX WDW	Fixed Window
G	Gas
GA	Gauge
GAL	Gallon
GALV	Galvanized
GB	Grade Beam
G.B.	Grade Break
GFI	Ground Fault Interruptor
GFCI	Ground Fault Circuit Interruptor
GI	Galvanized Iron
GL	Glazed (glass)
GND	Ground
GOVT	Government
GPH	Gallons Per Hour
GPM	Gallons Per Minute

GR	Grade
GR	Grate
GRAN	Granular
GRD	Ground
GRTG	Grating
GYP	Gypsum
GYP BD	Gypsum Board (sheetrock drywall)
HA	Hot Air
HB	Hose Bib
HC	Heating Coil
HD	Head
HDP	Head pressure
HDW	Hardware
HDWD	Hardwood
HEX	Hexagonal
HGT	Height
HM	Hollow Metal
HOR, HORIZ	Horizontal
HP	Horsepower
HR	Hour
HSE	House
HTR	Heater
HU	Humidifier
HV	Heating & Ventilation Unit(s)
HVAC	Heating, Ventilation & Air Conditioning
HW	Hot Water
HWH	Hot Water Heater
HWR	Hot Water Return
Hz	Hertz
I	I Beam
I	Iron
ID	Inside Diameter
IF	Inside Face
IMC	Intermediate Metal Conduit
IN	Inch
IN3	Cubic Inch

INC	Incorporate(d)
INCAND	Incandescent
INCL	Include(d)
INCR	Increaser
INS	Insulate
INT	Interior
INV	Invert (flow line)
I.P.	Iron Pipe
INV.	Invert
J	Junction
JAN	Janitor
JB	Junction Box
JB	Jamb
JP	Joist & Plank
JST	Joist
J.B.	Junction Box
K	Kilo (1000)
KAL	Kalamein
KD	Kiln Dried
Km	Kilometer
KP	Kick Plate
KVA	Kilo Volt Amperes
KW	Kilowatt
L	Left
L	Lumen
L	Length
LAB	Laboratory
LAM	Laminated
LAT	Lateral
LAU	Laundry
LAV	Lavatory
LB	Light Beam
LB	Pound
LBR	Lumber
L.C.	Length of Curve

LDG	Landing
LEV	Level
LF	Linoleum Floor
L.F.	Linear Foot
LG	Long
LH	Left Hand
LHR	Left Hand Reversed
LIN FT	Linear Feet
LL	Live Load
LOA	Length Overall
L&PP	Light & Power Panel
LP	Low Point
LPG	Liquified Petroleum Gas
LR	Living Room
LS	Lump Sum (Contract)
LT	Light (Pane of Glass)
LTH	Lath
LVR	Louver
LW	Light Weight
M	Bending Moment (force)
M	Meter
M	Thousand
MAGN	Magnesium
MAS	Masonry
MATL	Material
MAV	Manual Air Vent
MAX	Maximum
MBF	Thousand Board Feet
MBtu	Thousand British Thermal Units
MBR	Master Bedroom
M&D	Matched & Dressed
MDO	Medium Density Overlaid
MDP	Main Distribution Panel
MECH	Mechanical
MED	Medium
MET	Metal
MFD	Manufactured

MFG	Manufacture
MH	Manhour
MH	Manhole
MI	Mile
MIN	Minute
MIN	Minimum
MISC	Miscellaneous
MIX	Mixture
MK	Mark
ML	Metal Lath
MLF	Thousand Linear Feet
MDG	Molding
mm	Millimeter
MN	Main
MN	Mean (average)
Mo	Month
MO	Motor Operated
MOD	Modular
MOR	Mortar
MS	Manual Starter
MSF	Thousand Square Feet
MTD	Mounted
N	North
NATL	National
NBS	National Bureau of Standards
NEMA	National Electrical Manufacturers Assn
NF	Near Face
NFPA	National Fire Protection Association
NIC	Not In Contact
No	Number
NOM	Nominal
NOR	Normal
NRC	Noise Reduction Coefficient
NTS	Not To Scale
O to O	Out to Out
OA	Overall

OBS	Obscure
OC	On Center
O.C.	On Center
OCT	Octagonal
OD	Outside Diameter
OF	Outside Face
OFF	Office
OPNG	Opening
OPP	Opposite
OS	Outside
OUT	Outlet
OVHD	Overhead
OZ	Ounce
P	Plumb
PAR	Parallel
PART	Partition
PASS	Passageway
PAV	Paving
PCC	Point Of Compound Curve
P.C.C.	Portland Cement Concrete
PCR	Point of Curb Return
PCS	Pieces
PERF	Perforated
PERIM	Perimeter
PERP	Perpendicular
PG	Page
PH	Phase
PL	Pilot Light
PL	Plate
PL	Property Line
PLAS	Plaster
PLAT	Platform
PLBG	Plumbing
PLGL	Plate Glass
PL HT	Plate Height
PLMB	Plumb
PLY	Plywood

PNL	Panel
PPM	Parts Per Minute
PRCST	Precast
PR	Pair
PRC	Point of Reverse Curve
PREFAB	Prefabrication
PRESS	Pressure
PRI	Primary
PROP	Property
PROP	Proposed
PRV	Pressure Reducing (Regulating) Valve
PS	Pull Switch
PSI	Pounds Per Square Inch
PT	Point
P.T.	Pressure Treated
P.U.E.	Public Utility Easement
PVC	Polyvinyl Chloride (plastic water pipe)
P.V.I.	Point of Vertical Intersection
PVMT	Pavement
QT	Quarry Tile
QT	Quart
QTY	Quantity
R	Radius
R	Thermal Resistance
(R)	Radial
RA	Return Air
RAD	Radius
R.C.	Relative Compaction
RCP	Reinforced Concrete Pipe
RD	Round
RDM	Random
REC	Recessed
RECP	Receptacle
REF	Reference
REFL	Reflective
REFR	Refrigeration

REG	Register
REM	Remove
REINF	Reinforcing
REP	Repair
REQ	Required
RESIL	Resilient
RETG	Retaining
REV	Revision
RF	Roof
RFG	Roofing
RGH	Rough
RGH OPNG	Rough Opening
RH	Right Hand
RHR	Right Hand Reversed
RHW	Heat & Moisture Resistant Rubber
RIO	Rough-in Opening
RIV	Rivet
R/L	Random Lengths
R/W/L	Random Widths & Lengths
RM	Room
RO	Rough Opening
RPM	Revolutions Per Minute
RST	Rigid Steel Conduit
RT	Rubber Tile
RT	Right
RV	Relief Valve
RDW	Redwood
RWL	Rain Water Leader
s	Slope in Feet Per Foot
S	South
SA	Supply Air
SAD	Supply Air Defuser
SAF	Safety
SAN	Sanitary
SC	Scale
SC	Self Closing
SC	Solid Core

SCHED	Schedule
SCT	Sewer Clay Tile
SCUP	Scupper
SD	Storm Drain
SDMH	Storm Drain Manhole
SEC	Second
SEC	Secondary
SECT	Section
SER	Service
SEW	Sewer
SFCA	Square Feet of Form in contact with Concrete
SHLP	Shiplap
SHT.	Sheet
SHTG	Sheathing
SIM	Similar
SJ	Steel Joist
SM	Sheet Metal
SOV	Shut Off Valve
SPEC	Specification
Sq	Hundred Square Feet
SQ	Square
SQ FT	Square Foot
SQ YD	Square Yard
SR	Supply Register (air)
SS	Sanitary Sewer
SSB	Single Strength B Grade Glass
SSG	Single Strength Glass
SSMH	Sanitary Sewer Manhole
SST	Stainless Steel
STA.	Station
Std	Standard
STD	Stud
STL	Steel
STR	Straight
STRL	Structural
STRUC	Structural
SUB	Substitute

SUPP	Supplement
SUR	Surface
SUSP	Suspended
SV	Safety Valve
S/W, SW	Sidewalk
SWG	Standard Wire Gauge
SY	Square Yard
SYM	Symbol
SYM	Symmetrical
S2E	Surfaced on Two Edges
S2S	Surfaced on Two Sides
S4S	Surfaced on Four Sides
T	Thick
T	Threaded
T	Thermostat
TAN	Tangent
TB	Top of Beam
T&B	Top & Bottom
TBM	Temporary Bench Mark
TC	Top of Concrete
TC	Top of Curb
TEL	Telephone
TEMP	Temporary
TEMPL	Template
TERM	Terminal
TF	Top of Footing
TG	Top of Grate
T&G	Tongue & Groove
TH	Threshold
THK	Thickness
TM	Top of Masonry
TOF	Top of Footing
TP	Top of Pier
TRANS	Transformer
TS	Tensile Strength
TS	Time Switch
TS	Top of Slab

TV	Television
TW	Top of Wall
TYP	Typical
U	Up
UBC	Uniform Building Code
UG	Underground
UH	Unit Heater
UL	Underwriters Laboratory (electrical fixture specs)
UNEX	Unexcavated
UNF	Unfinished
UON	Unless Otherwise Noted
UR	Utility Room
USS	United States Standard
V	Valve
V	Vent
V	Volts
VAC	Vacuum Line
VD	Vent Duct
VAP PRF	Vapor proof
VC	Vitrified Clay
VEN	Veneer
VENT	Ventilate
VERT	Vertical
VEST	Vestibule
VI	Central Valve (as numbered)
VLF	Vertical Linear Feet
VOL	Volume
VP	Vitreous Pipe
VS	Vent Stack
VT	Vertical Tangent
VT&G	Vertical Tongue & Groove
VTR	Vent Through Roof
W	West
W	Width

W	Water
W/	With
WC	Water Closet
WD	Wood
WDW	Window
WF	Water Fountain
WGL	Wire Glass (reinforced)
WH	Water Heater
WHR	Watt Meter Hour
WI	Wrought Iron
WM	Water Meter
WP	Water Proof
WP	Weather Proof
WR BD	Weather Resistant Board
WS	Weather Stripping
WS	Water Service
WSCT	Wainscot
WT	Weight
WV	Water Valve
WWF	Welded Wire Fabric
X	By (multiplied by, as in 2 X 4)
X SECT	Cross Section (blueprint view)
XT	Temporary Marker
YD	Yard
YD3	Cubic Yard
YP	Yellow Pine
Z	Zinc

Index

About the Author

Jonathan F. Hutchings is a licensed building contractor and certified construction technologies consultant in California. He also serves on the California State Legal Advisory Board as an Expert Witness and consultant to the insurance industry. He is the author of several books, including ***CPM Construction Scheduler's Manual***, ***National Building Codes Handbook***, and ***OSHA Quick Guide For Residential Builders and Contractors***, all published by McGraw-Hill.